THE ENVIRONMENT TOTAL MASTERPLAN

Navigating architectural projects through planning and construction

Dr Graham Ford

Rethink

First published in Great Britain in 2023
by Rethink Press (www.rethinkpress.com)

© Copyright Graham Ford

All rights reserved. No part of this publication may be reproduced, stored in or introduced into a retrieval system, or transmitted, in any form, or by any means (electronic, mechanical, photocopying, recording or otherwise) without the prior written permission of the publisher.

The right of Graham Ford to be identified as the author of this work has been asserted by him in accordance with the Copyright, Designs and Patents Act 1988.

This book is sold subject to the condition that it shall not, by way of trade or otherwise, be lent, resold, hired out, or otherwise circulated without the publisher's prior consent in any form of binding or cover other than that in which it is published and without a similar condition including this condition being imposed on the subsequent purchaser.

*This book is dedicated to
Elena, Julia and Cecilia.*

Contents

Foreword	1
Introduction	5
The benefit of experience	6
The need to adapt	9
The roles architects use to help you	11
How this book will help	14

PART ONE   SETTING THE SCENE   17

1 How To Write Your Design Brief	19
What to include	21
Your value proposition	24
Accessibility	25
Integrating technology	26
Who can help?	27
Identifying risks	30
Testing the brief	32
Summary	35
2 A Proven Planning Process	39
What is planning?	40
What is permitted development?	42
A tried-and-tested method	43
Community Infrastructure Levy	46
Summary	51
3 Selecting A Contractor	55
Procurement	57
The background to design and build	59
The traditional route	63
Summary	64

4	**Typical Mistakes To Avoid**	67
	Choosing the wrong team	67
	Managing the construction yourself	68
	Tendering the job	69
	Not having enough design expertise	69
	Not allowing enough time	71
	Summary	71

PART TWO   SOLUTIONS 73

5	**Spatial Intelligence**	75
	What is spatial intelligence and why is it important?	76
	How do you discover your spatial intelligence?	77
	Confronting the spatial intelligence of another culture	78
	Summary	84
6	**Simplicity**	87
	Designing to achieve simplicity	88
	The pavilion	91
	Summary	100
7	**Precision**	103
	Tacit and explicit knowledge	104
	Accurate costing	108
	Summary	109
8	**Efficiency**	111
	Precision, efficiency and sustainability	113
	Prototypes and optimisation	116
	Building information modelling	119
	Information release and design changes	121
	Commissioning and post-occupancy evaluation	121
	Summary	123

9 Collaboration   127
    Roles to facilitate collaboration   128
    Collaboration with the contractor   129
    Summary   131

10 Sustainability   135
    A visionary masterplan   137
    Future-proofing your facility   139
    The impact of statutory requirements   143
    Embodied carbon   147
    Orientation   149
    Environmental assessment tools   150
    Summary   150

Conclusion   155
Acknowledgements   159
The Author   161

Foreword

My architectural career, which has spanned more than thirty years, has required me to embrace the ever-changing role of the architect within the design and construction industry. I have needed to be adaptable and resourceful, developing the procurement of architectural designs and our services to match varying construction techniques and project delivery methods proposed by the construction industry.

At an early stage in my progression from Architectural Technician to Managing Director of Skidmore, Owings and Merrill, and then into a consultancy role to architectural practices both large and small, I was taught that: *Every line is a decision.*

I interpreted this as meaning, before you physically undertake anything, such as the drawing of a simple line, you need to fully understand what 'the line' is.

Is it straight or curved, does it split into other lines, does it turn a corner and where does it change? What colour is the line? What thickness is the line? To create a form, which line needs to be in place first?

In this book, the analogy of 'the line' refers to 'the design and construction process'.

Changing the characteristics of the line as you go further down the design and construction process becomes more challenging, so making the most informed decisions in a timely manner is of great importance.

Graham Ford's book *The Total Environment Masterplan: Navigating architectural projects through planning and construction* sets out in clearly defined chapters some of the questions that clients, architects and contractors must ask to enable 'the line' to flow in many directions from paper to construction, from theory to practice.

It draws on his extensive experience of the design and construction industry and his desire to continue to improve the clarity of making informed decisions, which in turn helps deliver the designs of a project in the most effortless way.

It sets out different contractual relationship scenarios for varying project types and explains from the author's own experience what decisions were made in certain situations and how the outcomes affected the projects.

Graham's evaluation of the different types of contract forms provides the basis for selecting the most appropriate one for your project, and illustrates some of the benefits and pitfalls he has experienced either first or second hand.

Whether you are a new or established client with an estate to manage, an architect starting out on the challenging journey of your first project or an experienced building contractor, this book will bring benefits to your projects that you had not considered previously. It is a must-read for all in the construction professions aspiring to create exceptional projects, economically and in a timely way.

If you need an architect who can bring together a team or need to better understand the different design and construction routes, this book will certainly help you navigate the project through the planning and construction phases and give you a better insight into the choices you need to make to achieve a successful project.

Roger Whiteman, Dipl. Arch., ARB., RIBA was the Managing Director of American architects Skidmore, Owings and Merrill's London office before joining the British practice of RMJM. He led large major projects throughout the world, including Exchange House at Broadgate, offices at Stockley Park, office developments in Amsterdam, Warsaw and, more recently, Europe's tallest building, the Lakhta Centre in St Petersburg, Russia. He is now the Director of his own consultancy business **beside design**, which offers assistance to the profession to answer the challenging questions raised during the management of projects both large and small.

Introduction

Designing and constructing a building is a creative process. The resulting facility or development can transform the way your business works and how your tenants, students, members or employees feel about being a part of your organisation. It is your opportunity to make a significant contribution to our collective cultural heritage and to leave a positive legacy.

This book is for business owners, building owners, developers, facilities managers, finance directors, bursars and head teachers who want to undertake a building design project or review their estate and commission a masterplan for the future. It is relevant to those who wish to renovate an existing building, and it is useful for project managers who need the spatial skills of an architect at an early stage to develop their brief and test options on-site. If you work in the commercial, sports, hospitality or leisure sectors, whether you are experienced or not, this book will be useful in helping you to account to the investors and boards to whom you are responsible for the outcome and expenditure of your project. If you feel uneasy about any aspect of the development of your club, school, property or business; if you are concerned about your ability to attract new and better clients; if you need to read up on essential background details; or if you want to know how to make your building sustainable, this is the book for you.

Your focus is your customers' user experiences and how you can satisfy their wants and needs. Once the hard work of gaining permissions and getting the building constructed has been completed, you will want a smart asset that includes hardware (the building) and software (the digital services you offer). You may also provide food and beverages, leisure facilities, medical services, sports coaching and wellness. You will also want to make sure that each piece you develop fits into a strategy for your entire estate that includes infrastructure such as car parking, arrival spaces, drop off zones, energy generation, landscape, connecting spaces and water management strategies. These all need to be considered at an early stage to make sure they are integrated into the design.

You may be competing locally or nationally with other institutions or clubs that are continually upgrading their facilities. You may need more space to accommodate new activities, initiatives or courses you have developed that will bring further income. You will undoubtedly want to increase the revenue you take from the bar, from your dining facilities or from renting out space for external events. It is possible you will have requirements that appear to be at odds with each other, such as wanting to complete a construction project while keeping your venue operating at the same time. New facilities will help to create a story, build your brand and assist in developing a loyal following. You want customers to come to your building because of your values, your ethos, your style and the user experience you provide.

The benefit of experience

Your clarity about your project's outcomes and your deep understanding of the project from a business or strategic angle will be invaluable to your design team. You will rely on your architect and design team for advice on developing the brief, on the design, planning and construction stages, and on the budget. You will need to trust that your consultants are providing accurate information and leading you in the right direction to ensure your development runs smoothly.

As the client, you may be vulnerable to the risks of the construction industry and you may not be aware of some of the ways you can be overcharged or how things can go wrong. This is why you need advice on how best to set up and manage the project. The planning permission process is complex and costs a lot of money and time. Planning restrictions, especially in heritage environments and Royal and Regional parks, require an experienced team to prevent delays, costly mistakes and planning failure. You should allow your architect to bring in their network of experts to help guide you through this process.

The other big challenge you face is selecting the best procurement route for your project and how to choose a contractor to realise your vision. Do you opt for a traditional route where you and your design team are in control of the quality of the building? Or do you allow the contractor to manage the technical design phase on a design and build contract where all your hard work developing the design could be compromised?

Through years of the practical application of academic knowledge, adapted and refined in the field, at Graham Ford Architects (GFA) we have learned to design buildings, put together strong arguments for development to present to local authorities, and run a smooth build. Our principles and implementation strategies have been developed by collaborating with some of the UK's leading architects. The projects we have worked on recently include a masterplan for Leiths School of Food and Wine, the River Club in Surrey and a redevelopment of the Masonic Lodge in Harrow. We collaborated with other architects to provide consultancy to the London 2012 Olympic Games and design services to the Masdar Institute of Science Technology in the UAE.

Reflecting on all our built projects, I developed the simplicity, precision, efficiency, collaboration and sustainability (SPECS) principles. These principles are the ingredients that help my team guide projects and address any wrong thinking that could jeopardise their success. They ensure your development runs smoothly by reducing the risk of overspending, bad design, poor-performing contractors and programme overruns. They also help you and your design team to design a building that 'treads lightly on the planet' and is elegant and functional. In Part Two, I will go into each of these principles in some detail so you are clear how essential they are to a well-managed project.

A key event in my early years in practice was winning a scholarship in 2000 to travel the world and interview engineers and architects who were involved in sustainable design projects. Following this scholarship, I completed a master's degree in sustainability based on my field research. Many of the most sustainable projects in this book have involved retrofitting existing buildings. The Roundhouse in Camden Town was a conversion of an old railway repair depot into a theatre. The Heal's Development involved significant upgrades to several buildings, which included improved modern ventilation systems and thermal performance. These projects were important due to the substantial energy savings made when the buildings were repurposed. Retaining existing buildings also prevents unnecessary and wasteful demolition, retains our built heritage and helps promote low-carbon retrofit as a viable option.

The need to adapt

There are opportunities in the digitalisation of the architectural and engineering professions, artificial intelligence, building information modelling (BIM) and modern methods of construction (MMC) to make construction more efficient and sustainable. These technologies will help you to manage your facility more effectively, reduce energy consumption, provide a better experience for your customers and make your staff more productive.

Even with the assistance of technology, you and your team will need to carefully consider the 'gap' between what is on the design documents and what is practical to execute in the field. There will always be some risk associated with this, no matter how competent your designer is, as it is impossible to solve everything on paper. Your team will need a strategy for overcoming this gap. The way buildings are procured and the level of information you provide to the contractor before construction begins are crucial to the success of your project.

If you are undertaking a capital project, you will be managing budgets, risk and uncertainty as well as facing big trends that are impacting us all. We are in an age of rapid change where there are both huge challenges and unparalleled opportunities.

To understand these opportunities, we need to look at projects from the point of view of the natural world and consider future energy supply, climate change, population growth and loss of biodiversity. In the future, we will see leisure, hospitality, learning environments and workplaces merging, and design will move up the value chain as it becomes more critical than ever in attracting new customers to your building.

The pandemic of 2020/21 exposed our broken relationship with nature. Forests, freshwater systems, savannahs, oceans and the biodiversity within them provide us with clean air, water and food, and create buffers that are a natural protection from viruses. With urban sprawl and deforestation, we are removing these natural buffers and expanding the zones where wildlife comes into contact with humans and pandemics can emerge. Throughout history the design of our cities and infrastructure has been a response to diseases such as cholera and a search for better light and fresh air. It is now much more likely that your customers will demand better accommodation with exemplary modern ventilation systems than they have done in the past. Your facility will need to be transformed to meet these demands if you want to keep your clients safe and help improve the productivity of your workforce.

Climate change and loss of biodiversity are the context in which we work. We know that 42% of the carbon pumped into the atmosphere comes from our buildings.[1] There are profound economic, geopolitical and climate threats as a result of our

---

[1] UKGBC, *Climate Change: UKGBC's vision for a sustainable built environment is one that mitigates and adapts to climate change* (UKGBC, 2022), www.ukgbc.org/climate-change-2, accessed 30 October 2022

addiction to oil. Much of the world's supply of oil and gas comes from highly unstable parts of the world. Reducing our reliance on fossil fuel energy reduces our dependence on energy supplies from countries governed by petro-dictators. We are now in the middle of an energy transition where oil and gas producers will, in the near future, become energy companies supplying abundant, clean and reliable electrons from a variety of renewable sources. The clean energy sector is likely to be the next big technological revolution.

We need to dramatically reduce the energy demand required to run our buildings. Imagine if you could reduce your energy bills through better design and at the same time produce energy that you could use and export to the grid. The good news is that the payback times for renewable energy technology are continuously reducing, and awareness of the impact of humans on the environment is growing. We now understand that global warming is triggering irregular seasonal change, which disturbs natural systems. We are becoming much more aware that changes in land and sea use are a massive threat to biodiversity, loss of habitat and loss of species.

The roles architects use to help you

Your architect will have skills that help you with design, strategy and the ability to mediate between design and construction. They must be agile, a problem solver, able to resolve conflicts and they must have a wide general knowledge relating to design and construction. The architect, as the natural leader of the team, will ensure that the different engineering and design specialisms are allowed to shine both individually and collectively.

During my doctorate, I made the tacit knowledge of my practice explicit by identifying the different roles I adopted. These consisted of how I operated as an interpreter, advocate, covert architect, change maker and storyteller. What is unique about the roles is how I defined them, and the tools and techniques I developed to support them. I developed the storytelling role to help me design in a collaborative manner with mentors, colleagues, clients and planners. This approach helped us to refine our design work, navigate the planning system and resolve the various tensions created by the different agendas, such as development versus conservation or heritage, and energy conservation.

Following the global financial crisis of 2008, I adjusted to a new reality by working as a consultant on large design and build projects. This introduced me to a completely new way of working and new types of projects, including sporting venues. It was during these projects that I developed the five roles I had already identified to resolve the design and construction challenges I was confronted with, where there was sometimes a lack of strong principles underpinning the process. The origin of these roles is found in projects such as the Roundhouse in Camden Town, in the Heal's Development and in the many projects where I learned to interpret and adapt existing buildings.

The **interpreter** needs to understand your goals and strategic aims so these can be translated into the unfamiliar language of a building or masterplan. If your project involves adapting an existing building, your architect will interpret references in that building to past traditions and architectural history.

The **storyteller** develops the story of the project, which is told using the language of architecture to set the scene, introduce the players and characters, structure the storyline and provide background information and a plot. Storytelling supports design work and the submission of planning applications. The 'story' begins with the predicament of navigating the project in areas of the city that could be highly contested and where gaining planning permission may be difficult and complex – for example, green belts and conservation areas. The social, economic and environmental concerns of the project are rooted in the history and culture of the site, based on deep knowledge of how it is used. Your architect will develop and shape the narrative in workshops with local residents, politicians and government agencies to build a consensus of support and a shared understanding. The result of this approach is that when the project is presented formally to the local authority, they recognise their contribution, which reduces the risk of rejection.

The **advocate** is a custodian for the built environment and promotes sustainable outcomes and good design. The role also extends to being an advocate for our designs or the designs of others and communicating effectively with the construction team to ensure the design concept is not lost during development.

The **covert architect** acts as your trusted independent advisor. They can assist in the design of the building or support the contractor to resolve a design or construction issue on-site. No matter who pays the covert architect, they are your 'safety net', working for the good of the project and paying close attention to key challenges faced by the team. A covert architect could be commissioned to observe planning meetings and provide advice independent of the design architect. Their objectivity can help unlock design blockages where the design architect may not be able to see how to resolve objections from the planning department. The covert architect can see the project from a different point of view. In a construction setting, the covert architect is called in to help produce critical information to assist the builders, sometimes separate from the architect responsible for the technical design.

The **change maker** creates scenarios for the future to help future-proof projects. These scenarios discover trends and early warning signals so the client's operation can adapt to future challenges. For example, I designed the Hyde Park Pavilion using a demountable system anticipating that in the future this project may need to be adapted or relocated.

Throughout the book I will describe how I have used these roles in different projects to help me in my work as both a design architect and as a design manager.

How this book will help

This book will take you through the entire process of building – from writing the brief, the design development stages and how to maximise your chances of gaining planning permission, to the documents that are required for construction and how to select the right builder and contract for your project. The book will set out the nine design and construction stages through which your project will progress – feasibility, concepts, final proposals, preconstruction, negotiation and contract, construction and delivery, handover, post-occupancy evaluation – and show you how your design continues to develop until you have occupied the building. It will help you understand how to keep your project on track and what you need to do at each step to make sure design, budgets and timelines are under control.

The book is written in two parts. In the first part, I outline some of the challenges you may face when designing and constructing a building. These challenges include writing the brief, planning constraints, costs, managing and mitigating risks and selecting a procurement route. In the second part I discuss spatial intelligence and explain the principles we use to guide our projects through the design and construction phases to help ensure each stage progresses smoothly and you end up with a great new asset for your organisation.

Interspersed throughout the discussion are stories about clients who came to us for strategic advice at the beginning of a project to find out if a building or a renovation was the answer to their problem. There are also stories about clients who needed help accessing the necessary information for construction. In all cases, our clients benefited from our principles and nine-step process, which resulted in a smooth planning submission and construction journey. Each chapter ends with a checklist so you can easily assess what needs to be done.

Once you have read this book, you will be in a much stronger position to procure quality architecture that encompasses both tangible assets (the building) and intangible assets, such as your brand, your story and the user experience you provide. The book will help you understand that the design of a building, and the management of design information and construction activities, including quality control, must be joined up and fully integrated for a project to run smoothly.

PART ONE
SETTING THE SCENE

ONE

How To Write Your Design Brief

Writing a strategic brief is the starting point for all projects, and will be completed in the first phase: discovery. Like all aspects of a design project, it is an act of discovery, the first step in imagining the possibility of something new. A well-written design brief will set the scene so you can commence the project with a full understanding of what your objectives and requirements are and what type and size of building you need.

The brief is your chance to demand an original and exciting solution, to pour your creative energy into the process, and set the tone and the philosophy for the project. It will summarise your business, why you are successful, what values are important to you and how the project fits into your overall business strategy. You should examine your goals and needs in depth to create a detailed snapshot of the physical, emotional, financial and investment factors that are fundamental to the project. In this chapter, I will discuss the components of a good brief and who you can use to help you write it.

The most value comes early in the briefing process from distilling and communicating everyone's ideas and visions. Your architect or client advisor can help you write the brief by inviting the right people around the table to better define your objectives and consider them from multiple perspectives. Your draft brief distils your vision and draws together your strategic expectations and operational priorities. It allows fundraising work to begin, helps you negotiate with stakeholders and neighbours, builds political and public support and enables you to engage the best consultants possible. You may commission a design team during the feasibility stage to translate your vision into drawings and models to assist your communication with the board, the local authority and the local community.

Your brief should ask the design team to interpret the site and all the technical information relating to how your business operates. You will need to give the team a clear direction on budget, planning constraints, environmental standards, design priorities and aesthetic judgements. You may ask the design team to review your entire estate, including landscape and access, for example, and assess whether you should renovate, expand or build a new facility. You will want to know how the new buildings help to improve your total environment. This discussion will be based on the cultural value of your asset and the ease with which it can be adapted to meet your needs. You could end up concluding that, through changes to programming, organisation and management, the existing buildings can be modernised or extended, or that the project can be reduced in scope. Your brief will be updated and tested as the design develops during the preconstruction stages, and it will eventually help your design team give form to your aspirations.

What to include

To help you prepare for your first meeting with the architect, you should write down your aims and objectives, what you know and what you might not yet know, and an outline of what size you think your new facility should be. Does the scope include access, landscape, placemaking (public space, the adjacent streets and parks), energy or infrastructure? Do you need a masterplan that looks at the phased development of your entire estate over the next ten years? Are there any legal issues that might prevent you from building? These could include restrictive covenants or permission from freeholders.

You should put together as much data as you can on how your organisation works at the moment. This will include answers to questions such as:

- How many are you?
- How much does it cost to run your facility?
- How will a new facility help improve your staff's productivity and customer satisfaction?
- What works and what does not work with your existing facility?
- How much car parking do you need?
- How much growth are you anticipating and how will this affect roads and parking?

Your architect needs to listen to find out everything about your operation, what challenges you face, what you are most concerned about, what is coming up next that will help or hinder your business, and why the project is important to you. In some projects, the design needs to respond to many user groups. If this is the case, you may want your architect to engage in stakeholder mapping to understand who uses the building and for what purpose. It will be the designer's job, in collaboration with you, to decide which voices take priority and how that is translated into built form.

The brief will describe how you use your buildings and how any alterations or additions will work in relation to existing rooms. You do not need to get everything perfect on your first attempt, as the brief will evolve in response to the work the design team produces; you do need to understand the impact any new activities will have. You should outline who your customers or clients are, what kind of experiences you want them to have, their aspirations and needs and why they come to your venue. You should also include a list of your priorities, things that are negotiable and things that are non-negotiable. This is important as it will guide your design. A brief should state in high-level terms what the key technical requirements might be. These could include:

- Acoustics
- Robust materials to reduce maintenance costs
- Ease of cleaning both internally and externally
- Energy efficiency
- Future energy needs

- Capacity of the existing drainage
- Routes and upgrades of IT, gas and water
- Renovation and any heritage issues

One of the critical problems in any renovation project is how the new and existing spaces can be made to work effectively together through elegantly resolved solutions for circulation. It is important to consider specific spatial requirements and which rooms need to be close together (so-called key adjacencies). Your brief should raise and answer the following questions:

- How will vehicle movements for visitors, staff and members be organised?
- What is the increase in the number of people you are anticipating in the future?
- How will you manage deliveries and what will the frequency and size of these be?
- How will pedestrian and cycle access, and access to emergency vehicles, be managed and resolved?
- How will people access the landscape and how will this be used for sport and recreation?

When developing your brief, you should consider future changes and possible adaptations. For example, if your building is designed as a robust shell with non-load-bearing partitions, changes are relatively easy to make. You should discuss different ways of distributing heating and cooling to rooms that might change size and location over time, and consider what technology, such as heat pumps, you can take advantage of to support the ongoing decarbonisation of the grid. Finally, your brief should ask the design team to consider how the building engages with the public realm and makes a positive contribution to the community.

Your value proposition

Your value proposition is a statement of desired outcomes based on how the new building can support your business, generate revenue and drive demand for membership. A well-designed building will make your staff and customers healthier and more productive and it will future-proof your business.

Your value proposition will address the financial value of your project and other less tangible dimensions, including its environmental, social and cultural value. It is important to define what success looks like for each of these aspects and answer the following questions:

- What would a successful sustainable building mean to you?
- How do you define success?
- Would your energy bills need to be reduced to a certain level so payback times for investment in sustainable technologies are worth it?
- Are natural materials important to you?
- How important are your existing buildings and do they have significant heritage and cultural value?

It is crucial to clearly articulate the environmental, social and cultural values of a project to back up your ideas within your organisation and with stakeholders, and to support your planning application.

It is also critical in this early stage to have a budget in mind. This could be quite a daunting process and you may need some guidance on how to put one together. Once you have an idea of the size of your new building or extension, some basic rates per square metre can be used to get a feel for what the costs might be and what funding is available to help you.

Accessibility

One of the key considerations when writing your brief is making sure it contains information on how accessible your premises should be. You will need to identify the different groups that use your facilities so the design can accommodate them all. Some of the issues to consider include:

- How will your customers, members, visitors and staff travel to your venue?
- Where are the stations and how far away are they?
- Where will everyone park?
- What obstacles might people face inside your venue?
- Where are step-free access points and lifts located?
- How do your customers obtain refreshments?
- Where are the accessible toilets?

- How can you adapt your building to accommodate different users?

It is likely that you will have families with young children, visitors with complex needs, including those with sight loss or who are hard of hearing, visitors on the autistic spectrum and those with assistance animals. If your operation is large and you host several thousand people at your events, your team will need to run a customer service workshop with the design team so great solutions are considered up front from day one and integrated into the design.

Integrating technology

Part of your briefing should be to create a tech-enabled smart asset. This is important for all asset classes (residential, retail, leisure etc). If you have large events, technology will improve your customers' experience – for example, by enabling people to pre-book tickets and food and beverages. Technology enables the hardware, the software and the services to be fully integrated to provide a bespoke user experience.

To achieve this you will need to consider the following:

- Internet and data analytics to measure light and ventilation
- Medical services, including physiotherapy, hydrotherapy, doctors and massage therapy
- Access to leisure and hospitality (bar and catering)
- Sports coaching and fitness instructors

- Facilities for children and young adults, including separate changing rooms and rooms for gym, ballet, boxing, tennis, swimming etc

These separate teams must speak to each other during the briefing process. In a post-Covid world, it is more important than ever to provide a high-quality facility that meets your staff and customer needs and allays their fears – for example, through monitoring fresh air and providing good-quality light and a comfortable temperature.

Technology will help you with the future maintenance of your facility. This is particularly important with mechanical and electrical services so that any faults in the system can be tracked, located and fixed quickly. You should be open to the early adoption of new technology. This might involve fabricating part of your project in a factory to save time on-site and help minimise disruption to your current operations. An off-site manufacturing approach forces you to fix the design a lot earlier in the process. You may want your team to complete some initial research to confirm this is the best way to build the project.

Who can help?

If your project is small, like a tennis or cricket pavilion, then your architect can write your brief. Your architect will interview you and write down your requirements, which will take some of the pressure off you and ensure they have understood your needs correctly.

If you manage a large organisation with stakeholders, a board or school governors, it may be necessary to appoint a project 'owner' who reports on progress. Before your designers start work, it might be useful to consult with end users through surveys, focus groups and informal discussions.

If your project is complex and requires a lot of strategic thinking and/or interventions into an estate of buildings, you may decide to work with a client advisor, your preferred architect or one of your team to write the brief. A client advisor is an experienced architect and professional practitioner (but not the one designing the building) who works with your team. Your advisor or architect will consider the way your buildings are constructed. They should review the costs and benefits of sustainable heating and cooling systems and how you can mitigate the impacts of Covid and climate change on your operation.

One advantage of using an advisor is that they are independent of your design team and therefore add impartiality and challenge decisions, much like a nonexecutive director on a board would do. They should have highly developed diagnostic skills so that the advice they give is tailored to your specific needs. If you choose a design advisor who is up to date with their knowledge, they can advise on procurement, design quality, sustainability and BIM, and assist you in selecting the best consultants for the project. Working with a design advisor will lead to greater efficiency when the design team begin their work, as the project then has a clear direction.

If you are looking at several sites with a view to purchasing, you should consult your architect and get advice before you proceed. Your architect or advisor will undertake a feasibility study, which reviews the best location for the building against an initial brief that you have developed together. The site will be assessed from an environmental point of view, including for solar energy and wind power. It is also important to consider planning constraints, noise from aircraft or traffic, ground conditions – including any possible soil contamination – and how access will work for visitors and deliveries.

If you are thinking of purchasing and renovating an existing building, your architect can advise on any historic features that might need to be retained and what condition surveys you will need to complete. An environmental analysis will determine what thermal upgrades are required to meet current and future standards, and if you will need to upgrade mechanical, electrical and IT services. You will want to know the feasibility of adding new features and improving the way the various spaces might be connected with stairs, bridges or lifts etc.

On a large and complex project, you may decide to employ a project manager if you do not have a professional team in-house to assist you. The project manager is particularly useful if you need to commence building at an early stage or you want to construct the project in phases and tender packages of work, such as demolition or structure, to get things moving quickly. Project managers can help with the budget and programme, and with procuring design and construction services, managing the construction project and acting as contract administrators. They can advise on different types of contracts, such as design and build, traditional and construction management.

Identifying risks

A poor brief can result in you not getting the right design to support your vision, in rooms that are not the right size, or in you and your team not considering the following issues in enough detail:

- How your staff and members will access different areas of your facility
- Security and how you will limit access
- Good acoustics in classrooms and exercise spaces
- Compliance with fire regulations
- How to achieve natural ventilation solutions and keep noise and pollution levels at a minimum
- Where you need to mix natural ventilation with mechanical ventilation and extraction
- Solar screening to prevent overheating, especially in summer
- Other non-negotiable items specific to your operation, such as being able to fit large pieces of equipment into your building

If you don't consider utilities, these can take a long time to resolve with your service providers, holding up your project and making it difficult to replace or upgrade them in the future.

It is important that you communicate non-negotiable items, such as the impact of construction on your club or school activities. These items might include how long the construction programme will take and the importance of scheduling noisy work outside of term time or when it has the least impact. You will also need to consider whether temporary accommodation is required for pupils or staff.

The other key issue is procurement. Traditionally, buildings were procured following the completion of a set of working drawings by the design team. These working drawings were the basis of the contract and were detailed so that all the unknown issues were resolved, and quality and craft were embedded into the design. The downside of this approach was that it did not involve the contractor, so issues of buildability were sometimes not addressed early enough. Throughout this book, I will discuss procurement in more detail, including how different hybrid models can be used to help ensure quality is retained and construction logistics are built into the design.

A lot of money is invested in a new construction project, which is why the briefing process is so important. Everyone in the team must work hard to identify the risks up front and place them on the risk register so that they can be managed and mitigated as soon as possible. The financial risk is associated with the type of project that is being undertaken and where it is located – for example, in a conservation area or green belt. The risk register is a valuable document that helps you and your design team track how the risks are being mitigated.

When renovating an existing building or working close to existing structures, the risks are buried inside walls and under floors. One of the biggest challenges is upgrading services in historic buildings. Your team may need to commission invasive and noninvasive surveys to confirm the existing structure and how the new additions will connect to it. I was recently involved with a project where stability of an existing support wall for a railway bridge had not been properly assessed by the original design team, leading to redesign, delays and extra costs for the client.

Testing the brief

In the feasibility stage, your architect will test your brief with a design diagram (or what we call 'test fits') to determine what will fit on the site. For example, we help our clients by using readily available Ordnance Survey maps of the site in standard digital formats. We commence work by placing blocks representing different activities, such as changing rooms, tennis courts or gyms, on the site to help test what will fit. The resulting visual diagram helps our clients decide what rooms and activities they really need and how to combine and arrange them at a high-level view. This process allows us to develop the strategic brief and identify priorities.

After looking at the physical (site), statutory (building regulations, for example fire) and planning constraints, including impact on your neighbours, the diagram or test fits will be adjusted. This stage can be costed and the resulting feasibility document is critical to establishing whether your project is viable.

If you decide it is viable and you want to go to the local authority to discuss these ideas in a pre-application meeting, it will be necessary for your architect to work from a topographic survey of the site with levels and existing structures plotted accurately.

This phase is called concept design, where the test fit diagrams are further interrogated and developed into architectural drawings in three dimensions. This might include 2D drawings, plans and elevations or a 3D physical or computer model. It is worth noting that at each stage of the design, your brief is further developed and refined. The architectural drawings are reviewed and we go through a 'reverse briefing' process where we test to make sure your requirements are being met.

▶▶ **CASE STUDY: LEITHS SCHOOL OF FOOD AND WINE MASTERPLAN, KITCHEN AND DINING FACILITY** ◀◀

We helped Leiths School of Food and Wine develop a brief and a design for their premises in West London. The meetings were held with the managing director of the school, who wanted us to review renovation of the entire premises. Her aim was to improve productivity and make their facilities more attractive and functional for staff and students. The site is constrained and surrounded by residential houses, so any external changes were likely to be controversial.

We commenced work with a masterplan of the complex. Risk was identified and mitigated by engaging with a structural engineer to confirm which walls could be removed. The mechanical and electrical engineers established which services could be relocated so that the staff changing rooms and the ground floor kitchen could be expanded.

We also researched the planning constraints and how difficult it might be to expand the building at first and second floor levels.

As there were proposed changes on every floor, we quantified potential costs and clarified priorities in the first month so that the most important changes could be tackled first and scheduled during the summer break. One key issue was that the building had no spare electrical capacity. We conducted an audit of usage to establish exactly what capacity was available, and we explored how the capacity might be increased with renewable sources of energy or the construction of a new substation.

Once we had mapped all the structural and services issues, we were commissioned to commence design work for refurbishing the teaching kitchen and dining room on the second floor. The written brief was developed in collaboration with a project liaison officer, who interviewed staff and then confirmed the client's requirements and non-negotiable items.

The feasibility stage involved a detailed analysis of how the staff and clients would use the new teaching kitchen and dining area. Together with our client, we reviewed joinery, flooring, finishes, furniture design and dining configurations. We designed a bespoke table that fit the space and could be adjusted to match the number of guests in the room. We then translated all the information into architectural drawings and layouts for the contractor.

We discussed procurement with the client. Given the project's size and the specific operational requirements, we decided to proceed with a traditional contract and a full set of working drawings. We used a quantity surveyor to produce a cost plan so we knew the client was getting a fair price.

The negotiation with the contractor was important to build trust and rapport so we could work through the programme dates and issues of buildability together. The project was successfully completed in 2020.

LEITHS' INTERIOR

Summary

The brief is a creative and professional document that sets the tone and agenda for any new project. It establishes at the outset what you want to achieve and what is negotiable and non-negotiable. You may decide to engage your architect, a staff member who knows your operation, or a client advisor to help you develop and write your brief. Completing the feasibility phase will confirm the viability of the project and how it will fit with the strategic plan for your business.

When preparing your brief, you should be open to new design ideas, new ways of looking at the problem, and radical solutions. The team should review how technology can help you to build faster and more efficiently. Through this process, the project may take some unexpected turns that end up being transformational for your development.

If you are working with a client advisor and a project manager during the feasibility phase, there must be no overlaps in the duties that each consultant performs. The project manager should focus on management tasks, the management of risks, procurement and the budget. The client advisor should focus on the strategic brief, the selection of the design team, and any early stage feasibility work that might need to be completed before that team is appointed.

The feasibility phase of the project includes writing the brief and establishing a risk register with mitigation actions. At the completion of this phase, you will have fully costed test fit diagrams and enough information to know if the project is viable. You will be able to move confidently into the concept design phase.

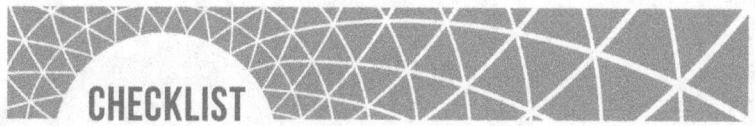

CHECKLIST

- ◯ Define your objectives, non-negotiable items and key technical requirements.

- ◯ Define the scope of the study you need to undertake. Does it include access, parking, energy, landscape and new buildings?

- ◯ Describe your customers, what they need and why they come to your facility.

- ◯ List your priorities – for example, quality, sustainability and accessibility.

- ◯ Make sure all your specialists, including catering, IT, management etc, are consulted when you are developing the brief.

- ◯ Commission a design advisor or architect to assist you and use their spatial knowledge to test how your brief will fit on the site.

- ◯ Commission invasive and noninvasive surveys of existing buildings to help the design team when these are needed.

ACTION STEP

To start thinking about your brief and the feasibility stage, and to make sure you are on the right track before you engage consultants, take our 'Feasibility study' scorecard.

You can find this at
www.grahamfordarchitects.com/scorecards

TWO
A Proven Planning Process

In this chapter, I will discuss planning and permitted development and take you through what is involved in preparing and submitting a planning application. Your application might undergo some serious scrutiny before, during and after its submission. Interested parties will review the proposed project's impact on neighbouring buildings, its contribution to placemaking and its impact on local infrastructure. This is the first major hurdle you might face on your project. The value of your site will significantly increase once planning is granted, so gaining planning permission is a major milestone.

When you reach this stage, you will already have completed a lot of work with your design team on developing your brief, interpreting the site and assessing different design options. You will need a proven process to make sure your application has the best possible chance of success, and you will need a team of professionals who can advise you on the economic, social and environmental impacts of your project. By comprehensively addressing these issues, you will make the process much smoother and end up with a better building or masterplan.

What is planning?

Planning policy is based on a set of principles established by the National Planning Policy Framework.[2] These policies are a guideline for local authorities to use when establishing the local plan, spatial development strategies or supplementary planning guidance. When you submit a planning application, your project will be reviewed against the policies contained within these documents.

The planning system is influenced by politics. Most large planning decisions will be voted on by elected councillors who will be influenced by their constituents. If the project is of national interest, it can be taken over or 'called in' by the Secretary of State for Ministry of Housing, Communities and Local Government. In London, projects can be reviewed by the mayor.

If your project is deemed to have a substantial impact on the local area, it may be reviewed by the local authority's design review panel, who will provide valuable feedback on your scheme. Your consultants will help manage the presentation to the review panel. The feedback from the panel is a 'material' consideration in the planning application process.

Most planning applications are not of national interest and few will be considered to have a substantial impact on the local area. If your project is for a school, university or sports club, it

---

2 Ministry of Housing, Communities and Local Government, *National Planning Policy Framework* (Ministry of Housing, Communities and Local Government, 2021), https://assets.publishing.service.gov.uk/government/uploads/system/uploads/attachment_data/file/1005759/NPPF_July_2021.pdf, accessed 30 October 2022

may be more straightforward. In this case, the outcome could be determined by the planner assigned to the project or, if more substantial, by the planning committee. Whatever the size and impact of your project, you will need to follow the principles outlined in this chapter.

If your project is rejected for any reason, you are entitled to appeal to the Planning Inspectorate, who will determine the outcome. The Planning Inspectorate is not influenced by constituents and local body politics and it is, unfortunately, sometimes necessary to appeal a decision by the local planning authority if you feel it has not been impartial.

Full planning permission

An application for full planning permission incorporates a design of the building, including the façade, external materials and landscaping. It will expire after three years unless work has commenced.

Outline planning and reserved matters

Before purchasing land or a building, you may wish to apply for outline planning to find out whether your proposal is likely to get planning consent. The scheme should be indicative and provide information about the building's impact on its surroundings. This will allow the council to make an informed decision on the development. Once outline planning permission is approved, details of the design – including appearance, access and landscaping – can be submitted to the local planning authority in a separate application, known as reserved matters.

What is permitted development?

You can perform certain types of work without needing to apply for planning permission. These are called 'permitted development rights'. These rights are granted by Parliament. In some designated areas, such as a conservation area or a national park, permitted development rights are more restricted. An article 4 direction, 'which enables the Secretary of State or the local planning authority to withdraw specified permitted development rights across a defined area',[3] can be introduced when the local council considers that the character of an area of acknowledged importance could be threatened. Where there are article 4 directions in place, you will have to submit a planning application for your project.

There are now permitted development rights mechanisms that can be drawn on to convert some high street uses into residential use. These permitted development rights contain 'classes' – tools which help developers achieve a change of use. We have successfully followed a prior approval, permitted development route to convert offices and restaurants into residential apartments in Kew Gardens and Twickenham.

If the building is listed, there are extra steps you need to go through during the planning process, and listed building consent must be applied for as well as planning permission. During the research and design phase, careful attention must be paid to the impact of removing parts of the existing building

---

3   GOV.UK, 'When is permission required?' (GOV.UK, 2014), https://www.gov.uk/guidance/when-is-permission-required#article4, accessed 4 November 2022

or adding new extensions. You will also need to consider what changes can be made to the way the rooms and circulation spaces are organised.

A tried-and-tested method

Without a tried-and-tested submission process, your planning application could be at risk. Rejection costs a lot of time and money. We have developed four key steps that have been thoroughly tested over hundreds of applications to create the best chance of success.

Step 1: Analysis

Analysis involves assessing planning risk so it can be mitigated as soon as possible. Each planning application should be interpreted in context, as solutions that work well in one area may not be acceptable to the local authority in another. All submissions should be designed to fit the particular circumstances of your site. This work will be commenced in the discovery phase when you are establishing the strategic brief.

In this work phase, we review the impact of any potential development on the neighbouring properties and the 'character' of the site. There will be several issues that you and your design team need to consider in depth, including the bulk and size of the development, its impact on the natural light available to neighbouring properties, and any increase in traffic and noise that may affect the area.

Because the planning system is influenced by precedents, to assess the chances of getting permission we need to know the details of the plot and examine how the local authority has treated similar applications in the past. We use our knowledge of both national and local planning guidance, including relevant local transport and environmental policies, to make decisions about what we think it will be possible to achieve on your site. It is important your team is thorough in the analysis of all policies to ensure something of key importance is not missed. Part of this step is deciding whether the precedents that have been established are reasonable or we should try and persuade local authorities to move away from their past judgements.

Step 2: Storytelling

Due to the complexity of the planning system, you and your team will need to make a compelling case for your project, using storytelling to help create a strong value proposition. Your design team needs to anticipate everything that the local planning authority will consider relevant.

The design is backed by evidence presented in a clear and structured way in a written document called the Design and Access Statement (DAS). A DAS is a comprehensive report outlining the process that has led to the development proposal. It presents a convincing argument as to why planning permission should be granted, and it is where your team can explain the value of your project in detail, including the economic, social and environmental benefits of the scheme. The more compelling your story, the better your chance of success.

Creating a strong story requires the architect to be a thinker, a feeler and a doer. The thinker analyses data; the feeler has empathy and understands psychology and the emotions behind your requirements and decisions; and the doer works out how to translate all the information into an action plan that solves anything that seems to be misaligned.

Step 3: Collaboration

The fate of your planning proposal is in the hands of planning officers and/or the planning committee. One of the potential advantages of a system as complex as planning is that through dialogue, empathy and negotiation it is possible to pick up hints, read between the lines and make the changes required to get your project over the line.

Consultation with the local authority is essential in all complex projects so that both you and your design team can fully understand what all the constraints are, and can benefit from the local authority planner's knowledge of the area and the local design guides. To successfully navigate planning, you should commission your architect to build a great team of professionals, including a planning consultant, engineers and a transport consultant, who have ideally worked together in the past.

Step 4: Redesign

Once we have collaborated with, listened to and received feedback from all the interested parties, particularly the local planning authority, your project may need to be reworked and transformed to accommodate the feedback. Only then can the revised scheme be submitted for planning.

Community Infrastructure Levy

When renovating or developing a new build, it is critical to bear in mind that once planning permission or listed building consent is obtained, the process is not complete. There will most likely be several conditions that need to be discharged before the project can commence on-site, and this process requires research, reports from experts and a submission through the planning portal to discharge the conditions. Time and resources need to be allocated to allow for this process. It is also important that you get advice from your consultants on what liabilities associated with the Community Infrastructure Levy (CIL) payments you might incur. These are charged by local authorities to support the costs of infrastructure.[4]

▶▶　　**CASE STUDY: HARROW MASONIC LODGE**　　◀◀

Analysis

The Masonic Lodge project was an extension to an existing mock Tudor pavilion located in a conservation area in the London borough of Brent. The brief included a large banqueting hall for 350 people, flexible meeting rooms with independent access, a new bar and a commercial kitchen. The analysis stage involved 'pulling' the existing building apart to reveal its underlying logic, designing

---

4   GOV.UK, 'Community Infrastructure Levy' (GOV.UK, 2014), www.gov.uk/guidance/community-infrastructure-levy, accessed 4 November 2022

ARTIST'S IMPRESSION

an extension on the western side of the existing building, landscaping and a new access route into the site. Our first task was to write a brief for the client and develop a design that fitted the 350 covers they required, supported by a new commercial kitchen. We also spent a lot of time researching the history of the site and the development of the centre. We gained detailed knowledge of the character of the area around our building, including all relevant precedents.

Storytelling

The storytelling stage commenced with the development of a value proposition based on how the project would support our client's business operation, employ local people, use local materials and engage with businesses in the area. The design focused on the future health and wellbeing of staff and visitors, and on the quality of spaces we could provide. We achieved this by opening up the building to natural light and views over the landscape, and providing exceptional internal air quality. The value proposition also extended to our client's environmental values and their desire to reduce the amount of energy required to heat and cool the building. This is predicted to reduce the centre's carbon emissions. We collaborated with our mechanical consultant, who helped us to deliver a sustainable design solution using natural ventilation.

Collaboration

The story developed through diplomacy and building rapport with the conservation officer and planner. By engaging with them during the pre- and post-application meetings, we understood their interpretation of the existing building and their views on how it might be transformed. Allowing the local authority to influence the narrative made them

partners in the design story, and the ongoing dialogue helped us negotiate a contemporary design response that worked for all parties. In this project there was tension between development and conservation, and between heritage and energy-saving strategies. It was our job to resolve these tensions.

Redesign

When the project became public during the planning process, different value systems conflicted with each other. Our predicament was to resolve the conflict between the client's brief for a large extension and the local authority's concern for the impact of the extension on the conservation area through increased traffic, noise from large events or the size of the proposed building. One of our roles was to understand and manage the agendas of both our client

ELEVATION OF SCHEME

and the local authority. The main challenge was convincing the local authority that any new structure would not harm the conservation area.

During the initial design phase, our client was reluctant to reduce the size of the proposed extension. To help them understand what was achievable on the site, we arranged a pre-application meeting with the local authority to openly debate the design, heritage and planning issues. These meetings with the planners and the conservation officer helped shape the design of the centre.

Following written feedback from the planners and the conservation officer, our redesign process included looking at many alternative scenarios, and brainstorming solutions. We reviewed the connections with the conservation area and redistributed the accommodation to reduce the bulk of the proposals. We gained agreement on the design approach from the conservation officer before our formal submission.

The mechanical engineer was pivotal in achieving our heritage aims; they kept the height of the link building as low as possible through a natural ventilation strategy that used windcatchers instead of large diameter ductwork. This reduced the visual impact of the extension on the existing pavilion, as required by the conservation officer from the council.

Conclusions

The project involved reinventing the old in a new combination of innovation and conservation that helped recontextualise the building to meet the client's needs. One of the key learnings from this project was the strong value proposition that encompassed social, economic and environmental values and which helped us argue the case for the development.

The four-stage journey was vital to planning success: it provided a structure to follow which helped with the engagement of different stakeholders. The resulting interactions with the client, the local authority, the design team and the collaborators all played a role in ensuring a successful outcome. Planning permission was granted in August 2021.

Summary

This chapter has discussed the planning process and the four critical stages to reducing planning risk and gaining permission for your development. It is important to understand the value of the project to you and your customers. The following questions will help you think about this more deeply:

- What is the economic value of your project?
- How will the project contribute to how your community work and interact together?

- How will the new facility improve your staff's efficiency and productivity?

- Has your project increased its environmental value by using less energy or enhancing your members' or pupils' wellbeing through more sustainable systems, such as natural ventilation?

- Do you believe the building might reduce crime rates in the local area, reduce social segregation or provide employment?

These considerations are paramount following the Covid pandemic, and the benefits of what you are proposing need to be clearly articulated.

Selecting a design team that can collaborate with the local authority is key to reducing the risk of having your application rejected, which costs a lot of money and time. As we have seen in the case study on the Harrow Masonic Lodge, collaboration with the planning authority at an early stage is critical in all large and complex projects. All interested parties will come to the table with different agendas, and it is the architect's job to make sure everyone is happy with the outcome. The project must meet your objectives, but it also needs to make a positive contribution to the community. Sustainability is important because it will reduce your ongoing energy costs, and because it demonstrates a commitment to quality and to lowering our collective carbon footprint. There may be conflicts in conservation areas between energy conservation and heritage, which your design team needs to resolve to achieve a positive outcome.

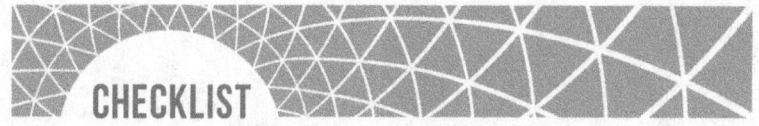

CHECKLIST

- ○ There are four steps you need to follow during the planning phase of your project: analysis, storytelling, collaboration and redesign.

- ○ Your team needs to advise what type of permission is best for your project: permitted development, outline planning, full planning permission.

- ○ Build a strong value proposition that will appeal to stakeholders, including the local authority and neighbours.

- ○ Establish what costs are involved in the application process, including the CIL payments.

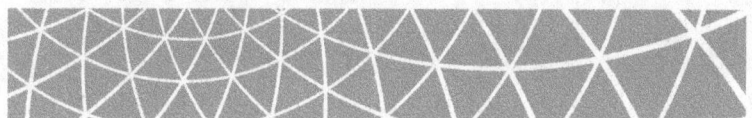

ACTION STEP

To start thinking about the planning stage, take our 'Do you need help completing a feasibility study to test your ideas?' scorecard.

You can find this at
www.grahamfordarchitects.com/scorecards

THREE
Selecting A Contractor

Apart from choosing your project manager and design team, the most important decision you will make is the selection of your contractor and the type of contract you will use to engage them. This decision needs to be made early, as working with the contractor will encourage innovation and contributions to discussions on buildability.

In this chapter, I will discuss the advantages and disadvantages of the two main types of contracts used in the UK: traditional and design and build. I will discuss why design and build became the procurement route of choice for public projects and what could go wrong. This is especially important following the Grenfell Tower tragedy. I have had direct negative experience of developing the technical design (including fire compliance issues) on transport and residential projects which were initially designed by other architects. I will also introduce a case study which demonstrates how my team delivered a complex renovation of a block of inner-city buildings. During my research for this book, I conducted interviews with leading architects who have influenced my practice. I will discuss the insights that these conversations revealed.

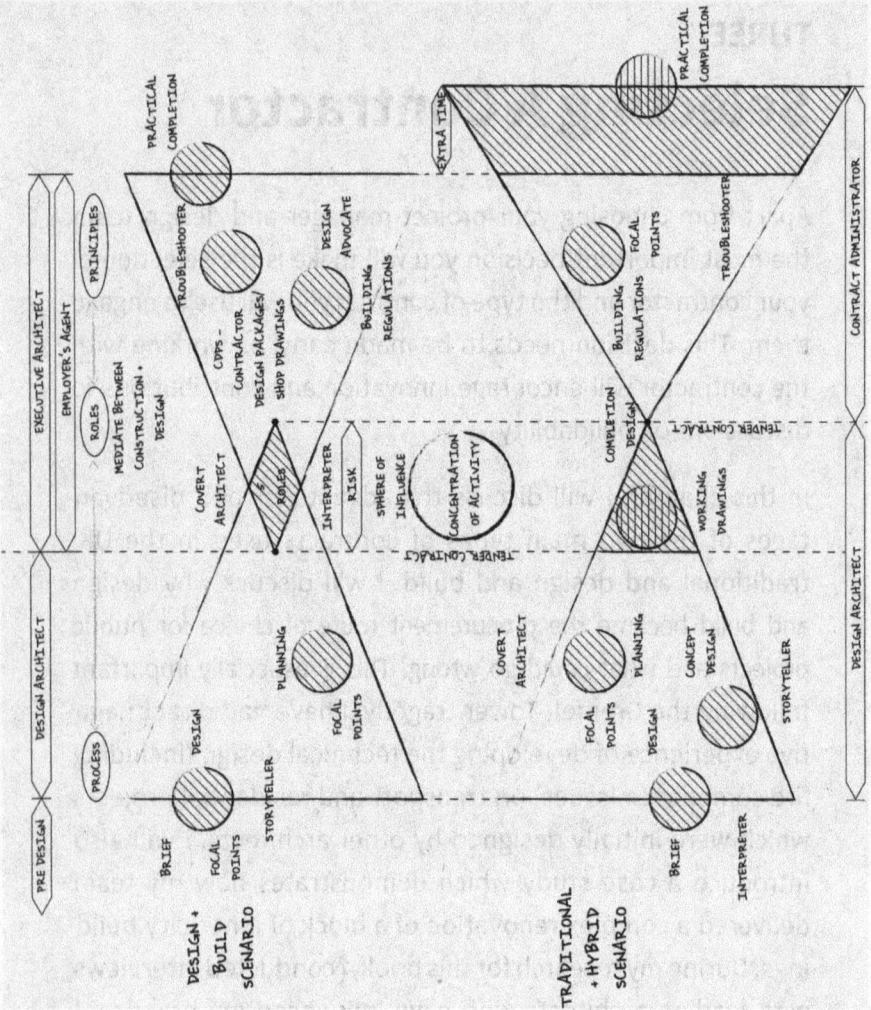

PROCUREMENT DIAGRAM

Procurement

Procurement is the way you and your team select the best contractual arrangement for your project. There are advantages and disadvantages to traditional and design and build. As we discussed in Chapter One, on smaller projects such as Leiths School of Food and Wine, where the scope is contained and quality is critical, a traditional contract may be the best option.

Where the building is larger, and early input from the contractor would add value to the design, a design and build procurement route could be considered. This route should be selected as a way of collaborating with a contractor, improving efficiency by taking advantage of their knowledge of construction. It is vital, no matter which procurement route you use, that your engineering and architectural team adequately assess the site and any risks associated with existing structures (bridges, retaining walls and buildings).

If you select a design and build procurement route, the documents which list the employer's requirements are prepared by your consultants. You will have control over the design elements that are developed in the early phases of the project. Once the contract is let, however, responsibility for design passes to the contractor to manage, and you will have no direct control. The contractor is responsible for the technical design, organisation, control and construction of the works. They can appoint their own consultants (such as an executive architect) or they can engage yours.

If a design and build procurement route is to be used, it should be structured so that a robust set of drawings and specifications is completed before the final contract is signed. We would recommend producing the following:

- A full set of planning drawings
- A specification
- Key technical details of areas that are important to you, such as the reception
- Full disclosure of the contractor's design team and their credentials
- A strong contract drawn up by your project manager / quantity surveyor that requires the contractor to attend design review sessions, develop the technical design, and build and install all the materials and components as detailed in the above documents

It is important to understand that you will need to pay for these drawings no matter which procurement route is used, so it is better to produce them early so the contractor is clear about your design intent.

These documents must ensure that no product substitutions will be accepted. Far too often, elements such as doors, cladding and curtain walling that have been specified by your design team are substituted for inferior systems with the excuse of nonavailability due to long lead times. It must be clear in the documents that this cannot be accepted. Your team must be in control over how the contractor will develop the technical design. As the client, you should understand the scope of work for the design team and what they are being asked to produce.

You should have regular reviews with the contractor and their team to assess progress and quality of work.

We also recommend considering a two-stage design and build process. The first stage is competitive and the contractor is selected based on preliminary costs and overheads. A programme for the design development and works is established and an outline overall cost is put together. In this first stage, many of the risks associated with the site, services and design can be clarified and mitigated before entering into a contract with the builder. There must also be an exit strategy in case negotiations fail. I have experience with this approach on the Heal's Development in Central London, which I will discuss in more detail later in the chapter.

The background to design and build

According to Richard Saxon, the publication of the Egan Report[5] in 1998 resulted in a change in government policy over purchasing buildings on a single responsibility basis, 'with the architect hired by the builder to ensure that the client has no risk from any gap between design and construction'.[6] These policies led to significant changes in the way architects worked. A central theme of the Egan Report was the need to improve productivity and efficiency through more integrated working.

---

5   Sir J Egan, *Rethinking Construction* (Department of Trade and Industry, 1998), www.constructingexcellence.org.uk/wp-content/uploads/2014/10/rethinking_construction_report.pdf, accessed 9 November 2022
6   RG Saxon, *The Future of the Architectural Profession: A question of values* (2006), www.saxoncbe.com/profession-values.pdf, accessed 30 October 2022; RG Saxon, *Growth Through BIM* (Construction Industry Council, 2013), www.cic.org.uk/shop/growth-through-bim, accessed November 2022

A specific issue relating to design and build is when design intent and continuity of information is lost during the post-planning phases of a project. Projects pass through the hands of multiple design teams and information is lost in translation between one team and the next. As Dame Judith Hackitt[7] has stated, the 'golden thread' in the design, construction and operation of buildings is where design information is effectively communicated from one party to another.

Following the 2008 global economic crisis, I transitioned from following the standard architect work stages to working on parts of much larger, more complex projects. This way of working involved more collaboration, working on the edge rather than from the centre, taking more risks and working in unknown areas of practice. While undertaking this work, I found – over twenty years after the Egan Report – that the industry is still fragmented, and design and build has failed to deliver the radical change in approach Egan was proposing. What has happened is a partial loss of the traditional role of the architect as a co-ordinator of information and designer of all the details. I discovered on these contracts that there are now endless discussions with the contractor's procurement team about the scope of the architect's work and the level of detailed drawings they require. In this preconstruction phase, I have found gaps in projects where critical design information had not been produced.

---

7   J Hackitt, *Building a Safer Future: Independent review of building regulations and fire safety* (HMSO, 2018), https://assets.publishing.service.gov.uk/government/uploads/system/uploads/attachment_data/file/707785/Building_a_Safer_Future_-_web.pdf, accessed 30 October 2022

CASE STUDY: RETROFIT OF EXISTING BUILDINGS, BLOOMSBURY, LONDON.

I was contracted to assist a large commercial architectural practice in developing the technical design for the retrofit of several buildings on Tottenham Court Road, Central London. One of the objectives was to upgrade the thermal performance of the buildings to future-proof the estate. The development consisted of art deco, early Victorian, and 1950s and 1960s office buildings around a central courtyard. The site is in the Bloomsbury Conservation Area and the estate is historically significant and was listed as Grade II in 1974.

The project involved upgrading the buildings to provide 39,000 square feet of modern office accommodation. The design connected the buildings with glass lifts, a glazed atrium, a new lobby and entry sequence, and a full renovation and upgrade of office space, including new mechanical and electrical systems.

I was employed as a consultant because of my specialist skills in the renovation of historic buildings. I was introduced to the project between the first and second stage tenders. My initial role was to interpret the design drawings with the project architect to find gaps and inconsistencies. I needed to understand the project through a forensic analysis of the information that had been produced for planning. I then organised the team to produce the technical design.

We collaborated closely with external engineering consultants who had a long history of involvement in the project. I worked as a design advocate to help the team develop consistently co-ordinated information that interpreted and delivered the design intent. I reviewed all aspects of the build, including the glazing around the internal glass walls, the display boxes, glazed meeting rooms and atrium.

One of my aims was to make sure the contractor's scope of work was well defined, and that they had robust, co-ordinated information that responded to the demanding heritage context. The project needed to be costed correctly before they were contracted to deliver the second phase, to ensure the quality the client required was allowed for.

During the second stage tender period, I helped the team develop packages of work that were tendered to the supply chain. I was then involved with the contractor in the selection of key subcontractors, such as the curtain wall and glazing subcontractor, to ensure the best team was engaged. The selection process involved a tender interview, a discussion of the design, identification of risks, costs and programme, and an assessment of their capability to complete the work based on their experience.

Once the project commenced on-site, I was engaged as the employer's agent to ensure the quality of the build matched the design documents. During the project, my role was not to manage the architect's team, but to give the architect confidence to lead the team and to advise on strategy for completing the work. As a mentor to the lead architect, I needed to understand what she wanted to achieve and then advise how this could be done.

The result was a Grade A refurbishment of an estate of buildings with a new entrance, a new glass atrium and a new circulation core to provide access to the upper levels. We replaced the plant at roof level, thermally upgraded the entire building fabric, including existing roof coverings and comfort cooling, and installed heating and low-energy lighting. The project was completed in 2013, with the investors achieving a significant uplift in rental income from the renovated building.

The traditional route

The design and build route may be best for larger projects such as stadia, but if your project is a small clubhouse or you are renovating changing rooms, classrooms or a small sports hall, you may decide to use a traditional procurement route. This involves your design team taking your project through all the stages to planning, tender and construction. The disadvantage of a traditional route is that the contractor may exploit missing information from the tender documents to claim extra costs. This is why, if you proceed down this route, you need to select an architect with a track record of delivering a technical design. If you have a quantity surveyor on board, you should consider negotiating directly with one contractor, as we did at Leiths School of Food and Wine.

Summary

It is important that you thoroughly weigh up the advantages and disadvantages of different procurement routes so that you can make the best choice based on the size and complexity of what you want to build. As a client, it is important that the design you have developed with your architect is protected so you end up with a quality end product.

The traditional route has the disadvantage of not involving the contractor until late in the process. This can be rectified by inviting your preferred contractor to become involved early and negotiating directly with them. The contractor will review logistical issues and assist in the selection of building systems. Too often, engineers and architects design buildings that need to be modified once the contractor becomes involved. Consider negotiating contracts with the help of an experienced quantity surveyor who can guide you on costs and advise on contracts.

If your project is large enough, you may be advised by your architect and project manager to enter into a design and build contract with a builder to take over the technical design role and engage a design team.

There is no advantage in transferring risk over to the contractor; it is far better to incentivise innovation and encourage performance and proactive problem-solving. There are many ways to do this, and the Heal's Development case study shows how early research with a contractor and invasive surveys can de-risk a project and enable the creation of a robust set of drawings before construction begins.

When you select the contractor, they must have a team you feel comfortable working with. They need to be willing to engage in an open conversation about innovative ways to build, save money and time, and reduce risk. We advocate using a traditional contract on simpler and smaller projects, and a two-stage design and build contract on larger projects to ensure that your design intent, which has been developed so carefully through the planning process, is protected during construction.

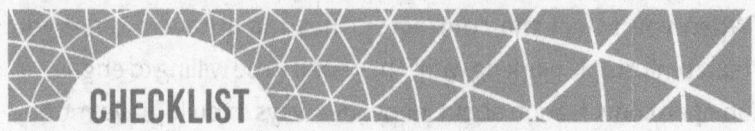

CHECKLIST

- ○ A traditional contract is where your design team develops all the planning information and completes the technical design.

- ○ A traditional procurement route is more suitable when the project has a budget under 5 million pounds and quality is paramount.

- ○ If you select a traditional procurement route, involve a contractor as soon as possible to discuss design and logistics.

- ○ If you are advised to use a design and build procurement route, consider using a two-stage process.

- ○ For a design and build contract, your team must provide robust design information and develop areas that are vitally important to you before going into contract with the builder.

- ○ On a design and build contract, ensure the contractor is obliged to install the materials and systems specified. Substitutions of specified products should not be accepted.

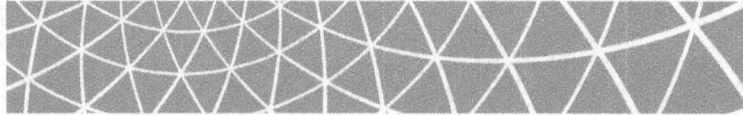

FOUR
Typical Mistakes To Avoid

This chapter will discuss what challenges you may face and the mistakes some clients have made in the past that you should be aware of. The most important step is to first select the right design team so they can advise you on which contractor to use. This will ensure your project runs smoothly through all the stages. Some clients are tempted to skip stages to save money, but this may lead to a lower-quality solution that could cause serious issues down the track. Your team need to be able to guide you through the concept and planning stages but they also need to have an overview of the entire construction process so that buildability is embedded into the design from the outset. The last thing you want is to have to go back to planning permission once it has been awarded and get approval for changes. This is expensive and takes up a lot of time.

Choosing the wrong team

The first big mistake clients could make is choosing a team from people they know, such as members of their club, parents from the school or family. The best option is to select your architect / client advisor and project manager based on their experience and skills and then work with the consultants they recommend. One poorly performing member of the team can cause delays and quality issues for the project.

The contractor is an important part of the team and everyone needs to approach the project with an open mind so you can benefit from mutual understanding. The objective is that the team learn from the different skill sets and ways of working in a collaborative setting where everyone has equal status. Lessons that have been learned from working together on other projects can work to your advantage. The best contractors will bring both practical building knowledge and innovative thinking to the table, which will save you time and money.

The risks are high in construction and margins are tight. Costs should be supervised by professional quantity surveyors so that the budget is managed and updated regularly. The team will help to identify risks early in the design process.

Managing the construction yourself

A complex project can end up a disaster if you try and oversee it yourself. No matter how good the construction documents from the engineers and the architect are, changes and clarifications will always be required. Unexpected things are sometimes encountered on construction sites and these need to be managed and costed. This process is critical to the contractor being paid fairly for changes to the construction documents.

It is also important that you don't pay more than you should for these changes. On most contracts this is the job of the contract administrator, who can be either the architect or the quantity surveyor. The contract administrator needs to be skilled at negotiation to resolve issues quickly and fairly. You need to trust their judgement and accept their final decision.

Regardless of the type of contract used, there are massive benefits from having an experienced and dedicated team working on your behalf during the construction phase on-site.

Tendering the job

One of the biggest mistakes you could make is tendering a project and then selecting a builder who has submitted a price that is significantly lower than all the other tenderers. In construction, it is well understood what it costs to build, and the materials and labour do not vary significantly from one builder to the next. A low price at the tender stage usually means the contractor will endeavour to exploit gaps in the engineering and architectural drawings to claim extra costs.

With some expert oversight from the quantity surveyor, all project costs can be accounted for. Negotiation with the contractor becomes a viable option, which means that the contractor feels part of the team and the management of the project becomes easier and the process smoother. If you're not comfortable with negotiation, your team can tender the job. There is, however, a cost associated with tendering a project, which you will need to consider before deciding how to proceed.

Not having enough design expertise

On a complex design project, you may be accountable to external funders, stakeholders and regulators, so you need high-quality project governance. In this case, it would be advisable to have a client advisor and/or project manager who can work with you to map out your brief and review your project strategy. Your client advisor should be experienced enough to help establish all the risks associated with your project. These are typically:

- The stability of existing structures, including adjacent walls and bridges
- Planning
- Services in the ground
- Contaminated ground
- Levels of flooding and water tables
- Party wall issues

On a design and build project, you should also consider working with an employer's agent. The employer's agent is similar to a design advocate and makes sure the quality of the design is reflected in the construction. This is done by assessing site works against the standards that are set out in the documents your design team have produced for you. If the original design team is not employed by the contractor, someone with skills of interpretation will assist in protecting the design intent by helping the contractor build what you have so carefully prepared.

Whether or not you are experienced in delivering projects, you will feel a great deal of responsibility for the project, which can be stressful. It is important you get the support and advice you need to make sure the construction phase runs smoothly and the design team does the heavy lifting for you.

Not allowing enough time

A key issue on many projects is not allowing enough time for innovation at the outset. It takes time for the best solutions to emerge, and accepting this will save time and money down the line. Another mistake clients sometimes make is not leaving enough time for construction. Everyone wants to make sure the project is completed as quickly as possible; however, you don't want to squeeze the process so much that the contractor is constantly rushing to complete the work and there is no time to ensure quality is maintained throughout the build. Your valuable asset must not be compromised because unrealistic timescales have been set.

Summary

No matter which procurement route you are using, your team must have thoroughly interrogated all the site issues and mitigated the risks. This includes investigations of ground bearing capacity, water table levels and services, and making the condition of existing structures known. Major problems can occur if a proper assessment of these issues has not been carried out and taken into account on the design.

If your business case supports a new or renovated building, you should select a team on the basis of their experience and ability to help you with the governance of the project and the design development. The cost of good advice pales in comparison with the cost of a building that does not work for you when complete, or the cost of disputes with contractors that arise out of misunderstandings associated with scope, quality and programme.

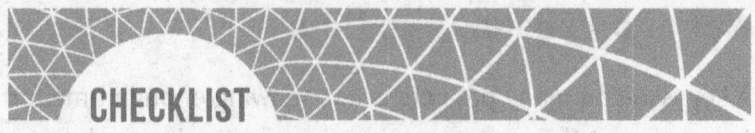

CHECKLIST

- ⭕ Choose your architect / client advisor and project manager based on their track record and skills and let them suggest their preferred engineering team and quantity surveyor.

- ⭕ Make sure your design team thoroughly investigates existing structures, services in the ground and water levels on your site.

- ⭕ Allow enough time for innovation at the beginning of a project.

- ⭕ Don't manage the project yourself; use a professional design team to help you.

- ⭕ Use an architect / design advisor to help with governance of the project, including the selection of the best design team possible.

- ⭕ Consider negotiating rather than tendering and use the expertise of your quantity surveyor.

- ⭕ Do not compromise quality by setting unrealistic completion dates for your build.

- ⭕ If you tender the project, don't select a contractor who submits a low price.

PART TWO
SOLUTIONS

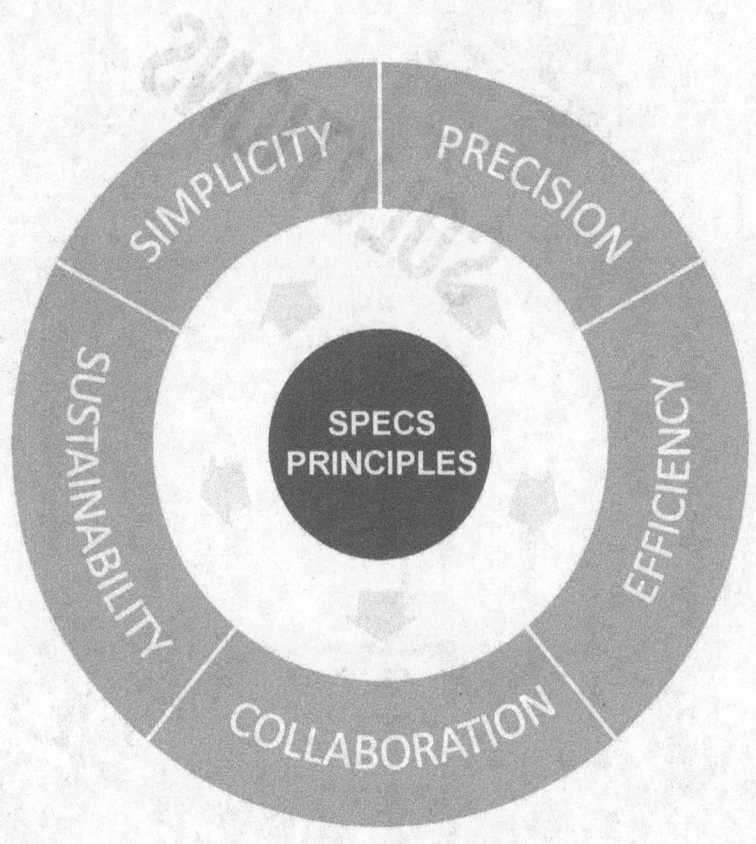

THE SPECS PRINCIPLES

FIVE
Spatial Intelligence

The design of your building is based on the spatial intelligence of your architect, a skill which informs their thought process. The more you understand the spatial design and its practical implications, the more knowledge you will have about the building and how it will work when complete. This knowledge allows you to make changes before it becomes too complex or difficult to do so.

In all design projects, there is some tension between the universal and the particular. It is likely your architect will build on the strong traditions associated with the place where your project is located. Architects are influenced by the artists and architects who have worked in your region over many decades, but they may also draw inspiration from other spatial and cultural traditions. Every architect and artist will build their own tradition as a synthesis of many influences.

The spatial intelligence of your architect will also help you in the early briefing stage. A project manager will help you organise the overall programme and assist with procurement, but you will need someone with spatial skills to work out how to organise your site, where to place new buildings and how to extend existing buildings.

You must always be upfront and clear if you don't understand the drawings – building plans are complex and not easy to read. It is important in the early briefing phase that your questions are addressed so your architect can prepare 3D images to help you understand the design. If you do not fully understand the design, the risk is that during construction you suddenly realise what is being proposed and want to change it. This normally causes huge difficulty with the build and can cost significant amounts of money.

What is spatial intelligence and why is it important?

When you engage your architect, you will work with someone who has both tacit and explicit knowledge. Tacit knowledge is intangible and has been embedded over time. Information, techniques and methods become tacit knowledge that is used to overcome new situations and challenges. The architect's tacit knowledge is know-how and memory, but first and foremost it is spatial intelligence.

Spatial intelligence was first codified by Howard Gardner,[8] a professor at Harvard University. Gardner believes that it is one of the eight intelligences: linguistic, logical-mathematical, musical, bodily-kinaesthetic, spatial-visual, nature smart intelligence, interpersonal and intrapersonal. Designers marshal their intelligence, especially their spatial intelligence, to construct the mental space within which they practise. When the artist or architect is conscious of what is motivating them, they are beginning to understand the archive of images that they have collected. These archived images from the past determine our interpretation of the present and enable us to visualise the future.

---

8   H Gardner, *Multiple Intelligences: The theory in practice* (Basic Books, 1993)

Explicit knowledge is tangible and rational. It is knowledge of local planning rules and the building regulations and information about the site where your building will be placed. It also includes information on your built assets and how your company operates, its culture and its organisational structure. The zone where tacit knowledge and explicit knowledge overlap is where the design for your building will develop.

If you select your architect based only on the criteria of delivery, without considering their capability for creative and innovative design, then you may end up with a mediocre building. For you, this might mean a design with poor-quality natural light and ventilation or a project that neither lifts the spirits nor attracts more clientele. The resulting building may not support your vision, mission or brand values, and the functional spaces in the building – such as plant, ducts and storage – may be in the wrong places and prevent great views out or effective communication and movement between different parts of your building or estate.

How do you discover your spatial intelligence?

Our spatial intelligence is formed from a collection of memories of buildings in both urban and rural landscapes from our past. My archive of images that forms my mental space includes the houses, churches and schools in the city I grew up in on the Canterbury Plains in the South Island of New Zealand. These buildings were a mixture of Californian bungalows, timber villas from the Victorian period and cathedrals based on the baroque architecture of Rome. The simple forms and spaces of the modernist houses with exposed blockwork were designed by the architects Warren and Mahoney. They were inspired by international influences from Scandinavia and the UK and were a radical departure from the Victorian villa.

While writing this book, I have mined my mental space and uncovered some of my fascinations and the images of art and architecture that motivate me. The holiday houses where I spent my summers were simple timber structures with an open connection to nature. These memories have become part of a collection of images that helped form my spatial intelligence and inform the buildings I design today.

You will have a mixture of the eight intelligences that have been identified by Howard Gardner. Your spatial intelligence will come into play with your vision of the building and the spaces you would like to create. The more you can describe and explain what images appeal to you and what motivates you, the more likely you are to get the building you want. When the architect responds to your brief with architectural designs, models and drawings, you will have a reaction to the spaces, the form and the shape of the proposed building. This is part of the creative process and you should be ready to enjoy the debates, brainstorming and differences of opinion which are all part of developing a design.

Confronting the spatial intelligence of another culture

When I moved to Europe in 2001, I came with a mental map of the openness of the New Zealand landscape – each house on a plot with front, back and side yards set in a gridded city full of parks and rivers. In London, I adapted to a different urban structure of Georgian and Victorian terraced housing mixed with a healthy distribution of urban squares with unique character. These terraced houses are made up of narrow plots, brick party walls and brick façades. I was challenged by these new spatial histories, which were dramatically different from the timber-framed villas and bungalows of New Zealand. I learned to manipulate and transform this stock of buildings.

London's public realm can be discovered in its parks and squares, whereas in most European cities we find the public realm in large boulevards or piazzas or even by walking through the porticos that cover the streets. These differences tell us something about how community has been organised. History is made visible in our cities, towns and landscapes. An example of this is the renovation of the Camden Roundhouse in London.

I was responsible for the renovation of the Roundhouse Theatre for over three years from 2004 to 2006 for the scheme architect John McAslan and Partners. During this period, I gained a wealth of knowledge of materials and construction techniques and a deep knowledge of the structure and geometry of the existing building.

This building is a good example of how a project adds to an architect's archive of mental images and spatial intelligence.

In the UK, there is a long tradition of technological know-how and skills in material assembly that were developed by engineers, designers and fabricators during Britain's role at the forefront of the Industrial Revolution. There was a willingness to experiment, evident in Joseph Paxton's glasshouses at Chatsworth and his Crystal Palace in Hyde Park, built in 1851. Paxton designed large structures using prefabricated components built at great speed to meet the programmatic requirements of a large exhibition space.

The Roundhouse Theatre, constructed in 1841, is part of this tradition and an example of the adoption of advanced engineering technology. It was used as a locomotive repair depot and was part of the second development phase of the first intercity railway in the world. The Victorians built a magnificent circular volume formed of brick with a roof supported by timber rafters and a cast-iron structure constructed with a high level of craft. They had both ambition and money to spend and left a significant legacy. The space has an emotional power that is enhanced by the materials, the volume and the way shafts of light enter through the roof and create a mood inside the building. The process for redeveloping this building involved redesigning, rethinking, reinventing, reinterpretation and relooking at everything in a new combination of conservation and innovation.

In the next case study, I will make explicit how I have used my spatial intelligence to redesign buildings, solve problems, collaborate, and help design and deliver a new sports club.

▶▶ CASE STUDY: SURREY SPORTS CLUB MASTERPLAN ◀◀

This project was for a client who owns a sports club in Surrey, close to London. The club comprises several buildings within a large landscape and there is a car park for members. The club is in Metropolitan Open Land, with the Hogsmill River on the southern boundary connecting it to a number of adjacent nature sites. The landscape is open grassland with some scattered areas of broad-leaved woodland. We anticipated that planning would be challenging and the biggest risk for the project, given its location in the green belt.

The brief from our client was to improve the club's facilities and grow the membership by attracting families. It was important to look at the entire site and anticipate how changes to buildings, car parking and landscape might be made over time.

On the site, there was a historic house where the gym was located, a swimming pool and several existing buildings of low-quality construction which held the changing rooms and smaller exercise areas. The client wanted to move the gym from the old house, where the spaces were small and unsuitable, into a larger space. The old house would then be converted into a restaurant.

The club had two major assets that needed to be improved. The enclosure for the swimming pool needed to be replaced and the tennis courts were in poor condition. We looked at several options of where to locate the new gym, tennis and padel tennis, and how to connect these activities to the changing rooms and restaurant. We designed spaces where young children and teenagers could take part in gymnastics, ballet, martial arts, tennis and swimming.

As planning was critical, before we began to develop the strategic brief we worked with our landscape architects to put together some sketches showing ideas for the site. We looked at access, car parking, the way the buildings related to each other and to the landscape. Our sketches showed ideas about how we could reconnect the fragments (buildings) on the site, enhance the existing buildings and work with the landscape to create a place for nature to develop. These sketches were then discussed with the local authority. This was a fast and effective way of getting informal feedback.

Once we had a positive response from the local planning authority, we had more confidence to proceed with developing the strategic brief. At the same time, we developed test fits on the site. Our spatial diagrams and the written strategic brief informed each other and helped the client clarify what was possible and where different activities could be located. We discussed our initial thoughts with our planning consultant and his feedback helped inform the development of the different options.

Once the client was happy with the design, we moved into the concept design phase with more developed drawings.

We provided information on building scale, materials and architecture so the planners could assess the impact of the proposals on the green belt. The project manager wanted to complete a pre-application meeting as soon as possible so we could receive written feedback on our ideas. This provided the client with some assurance of what the planners would accept.

The tension in the project was between the client's ambition to extend, improve and develop the club and the local authority's role of preventing any possible harm to the green belt through inappropriate development. Our job was to reconcile these two agendas.

We produced a design statement that set out our story based on the project's social, economic and environmental values. Our client was keen to continue public access to the site and develop the social infrastructure (a large community hall) that already existed. We investigated how we could provide access for local children and teenagers to use the new facilities for sports and cultural activities. The project's aim was to enhance and improve the buildings that already existed and reduce the amount of energy that the new facilities use.

The design benefited from the fact that the site is unpolluted, in green space and separated from the local suburban area. We could therefore use the wind from the south-west during summer to help with natural ventilation. The buildings were designed to maximise light to the facility and at the same time prevent overheating from the west during peak periods. At the time of writing, our masterplan was being presented to the local authorities through the pre-application process and we will be discussing our proposals with them. This collaboration phase will be followed by the redesign phase where we incorporate their comments into the final masterplan and building designs for formal planning submission.

Summary

Spatial intelligence enables your architect to access their interior world and tacit knowledge to discover spatial solutions for your project. These solutions are specific to you and the local environment and the result will be an assemblage of rooms and spaces which have volume, texture, light, detail and acoustic quality. Your building will be transformed and have a vitality that enhances the experiences and lives of those who occupy it.

The more you understand your own spatial intelligence, the better you will be able to communicate your vision and objectives and the quality of spaces that you need. Your architect's spatial intelligence can be used at the beginning of a project to help with your strategic brief. It is essential to test your ideas and brief using spatial diagrams. In this way, you can easily visualise what is possible before you start investing too much money and time into the design of individual buildings.

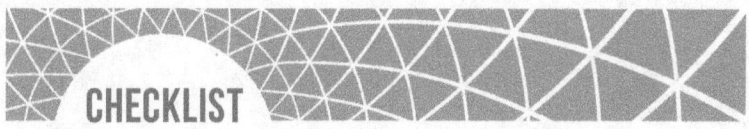

CHECKLIST

- ○ Select an architect whose team includes design and planning expertise and who can deliver a technical design.

- ○ Use the spatial design skills of your architect to help you with the strategic brief and to test what building or extension goes where on your site.

- ○ Be aware of your own spatial intelligence to help you understand what is being proposed.

- ○ Spatial intelligence is the foundation of your architect's thought process.

- ○ Design development continues throughout the technical design process.

- ○ The executive architect on a design and build project can significantly enhance your project with their spatial intelligence skills.

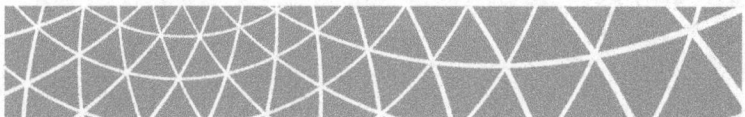

SIX
Simplicity

The next five chapters will go into more detail about the SPECS principles I introduced at the beginning of the book. The first of these is simplicity, which starts with your aims and objectives for the project. The clearer your goals, the more successfully your design team will be able to transform what you need into architecture.

Simplicity is an important design concept. It is about distilling the design down to the essence of what is necessary. It is about understanding how the elements in the design work to make a great building and offer a great experience for users. For example, the walls the architect places in different locations might channel wind into the building; provide privacy, refuge and structure; frame a view; or capture the sun at certain times of the day.

To achieve simplicity, your design team will work hard to refine the building diagram over time and test it against your goals and the detail in your brief. This principle is critical during the early stages of your project. In a good design, you must get the diagram of your building right. Once the architect has achieved this, the design process will flow.

The diagrams can usually be drawn on one sheet of paper. The initial diagram is a larger site diagram that shows how movement through the site, landscape and buildings work together to create new spaces that enhance the total environment.

Our diagrams take things apart – they are forensic. They show the mechanics of how landscapes and buildings work, how the engine works, not what it looks like. The diagram of the building is a search for a solution that works both functionally, structurally and environmentally.

The building diagram will encapsulate how people move through your building, how daylight enters through windows and doors, how your building will be ventilated. The complexity of the building's operations, such as security, air supply and servicing, must be considered in this diagram and integrated into the overall strategy. The result is a comfortable and ergonomically designed space that meets your needs and those of your customers. If your diagram is wrong then your building will be wrong, which means you have a much bigger problem because buildings are expensive to change.

Designing to achieve simplicity

During the feasibility phase on a smaller project like a pavilion or a small school building, the architect moves through the following steps: research, test, review and rework. This lays the groundwork for establishing the principles of how the site might be organised. The design diagram will normally become clear to everyone in the concept design phase.

The **research stage** involves streetscape studies, studies of adjacent squares, site access, views and reviewing the historical development of the site, including any previous planning and design studies of the area. To fully understand a site your team will need information on wildlife, ecology and biodiversity. If your project is a building for sport, your architect will review all the current requirements of the sporting bodies involved. This stage will include precedent studies where your design is benchmarked against buildings of the same type. The research stage overlaps with the analysis step, which is part of our four steps of planning.

The **test stage** is where your goals, vision and brief are translated into diagrams and models to help qualify your requirements. As discussed in Chapter One, your architect will complete test fits of your brief (the spaces you need, such as teaching spaces, changing rooms and meeting rooms) on the site. They will use design guides to get a feel for how the new design will accommodate the future number of users, how the design relates to existing structures, how the spaces connect and how the circulation both inside and around the outside of the building might work.

It is important that you identify your site boundaries in the briefing stage. The study area should extend them (conceptually) as far as necessary so you can consider the impacts of landscape, water, flooding and infrastructure on your site. If your building is for sport, your design team will review topography, in particular the placement of spectator viewing above the field of play. Other issues your design team will assess include:

- Wind and sun location in relation to spectators
- Playing field location in relation to pavilions
- Optimal orientation of field of play
- Scoreboard location
- Relationship of gym and swimming pool to changing rooms and medical rooms

- Location of outdoor nets and warm-up areas
- The extent of lighting for both the field of play and the car park and pavilion
- How the visitors, staff, pupils, parents and members arrive
- Public transport options and parking facilities, including disabled parking
- Location and parking for deliveries and access for emergency vehicles
- Urban and landscape studies to see how the proposals relate to both the built and natural context
- The management of flows of water (through storm water infrastructure) around the site to prevent flooding and pollution caused by excessive run-off
- The best way to integrate existing and historic buildings into your development strategy

If you are adapting an existing building or have adjacent buildings on your site, a surveyor will complete a measured survey so that when your architect starts sketching they have accurate information. This test stage will help your team to review options between different spatial solutions and then evaluate these against your goals, brief and subjective design criteria.

The **review and rework stages** involve a design workshop with your team where the initial ideas are explored, reviewed and then reworked following feedback. Once this phase is completed, you will receive a set of plans that will be essential for business case development. Once costs have been reviewed and approved, the project will be developed in the concept design stage.

Projects like sports pavilions or leisure centres can be great buildings on relatively modest budgets; all that is required is an innovative mindset. Costs can be controlled by the level of finish you choose and by keeping the form or shape of the building simple. You may, in discussion with your design team, explore different building forms or shapes if these add value. Different forms can yield more interesting spaces, provide better views or light, help ventilate the building or make your building more iconic, thus enhancing your brand.

The pavilion

Pavilions are important to the history of design in the twentieth century. They have been used to test composition, space, materials and ideas. They provide a great place for architects to learn how to develop a simple but clear diagram of how the building is going to work. For example, the concept for the Hyde Park Boating Pavilion was rational and clear with a retail space that opened on to the water. The ticket and management offices were located on the western side of the building with a large area for queuing and collecting life jackets before visitors stepped onto the jetty and into their boat. The staff kitchen and changing areas were located on the eastern end of the building. The pavilion feels light in the landscape as it was designed with clerestory windows that separate the roof from the timber base. The pavilion is a manufactured industrial product that provides flexible spaces and is demountable.

HYDE PARK BOATING PAVILION

The appeal of pavilions for me is associated with their informality, the idea of camping and being close to nature, and the simplicity of the plan layouts designed without corridors and formal entry spaces. The open structure of pavilions can be programmed in different ways. For example, the Hyde Park Boating Pavilion could be used as a classroom, changing rooms or a weekend holiday house. Pavilions are designed to be permeable to sun and wind. The idea of adaptability through modularity (adding on or removing bays, for example) and demountability is important in the design of pavilions. These ideas influenced a pavilion I designed in the Lee Valley.

CASE STUDY: STONEBRIDGE LOCK

Introduction

My client was the owner of Bluebird Boats. We had worked together on the design of the Hyde Park Boating Pavilion and we had an established working relationship. The project involved designing a new boating pavilion on existing hardstanding on the banks of the Lee Navigation in the Lee Valley Regional Park, East London. The site is separated from the suburb of Tottenham by road and rail infrastructure to the west. To the east, there is significant infrastructure, including reservoirs, wetlands and flood relief channels. Due to the small scale of this project, the design diagram become clear at the end of the feasibility study.

STONEBRIDGE LOCK MODEL

The project brief

The brief for the new building was to contain boat storage, staff facilities, a workspace, the ticket office and a large exhibition space for use by the community. My client needed enough design information to get the scheme costed and prepare a business plan to determine if the project was viable. I was asked to prepare a pre-application submission to the local council as the project was located within the green belt.

Research and interpret

There are several existing buildings on the site, including a café with a storage facility, a lock that is used by canal boats, a road for access to the local cabins and a car park used by walkers and birdwatchers.

My research involved assembling a relevant bibliography of books about the Lee Valley, old design reports, council documents and photographs that revealed how people once worked and played in the area. This helped me to link the physical characteristics of the river valley to the industrial heritage, which included pumping stations, locks, canals and reservoirs that formed a strong presence in the landscape. It also revealed where nature, history, community and culture converged and provided opportunities for architectural intervention through a new way of seeing the ordinary and valuing the industrial past. The way the landscape supported the physical and psychological needs of the community became part of the backstory.

I also engaged in conversations and meetings with various groups, including planners, activists, locals and the Canal and River Trust. The conversations I had with local people were instrumental in my understanding of their preferences, values, interests and concerns. These concerns included the impacts of flooding, fire and humans on the local migratory bird population. Our conversations also helped me to understand local planning policy, the political landscape with the council, and the tensions between the boating community and other groups that frequented the site. I discovered that the site was a matrix of contested space: on one side the complex management of the land by different government agencies and on the other illegal occupation by different groups.

Test

This phase involved testing the client's brief against a diagram and responding to the characteristics of the site. The design I developed opened out to the canal, as the focus of the building was directed towards activities on the water. I designed the building as two pavilions with a large open space between them and a roof that 'floated' over the top so that it would feel like the building was part of its natural setting.

The new building has a connection with the existing buildings through our proposed upgrade of the public realm. There is a raised platform for the accommodation block to take into account possible future flooding

The Total Environment Masterplan

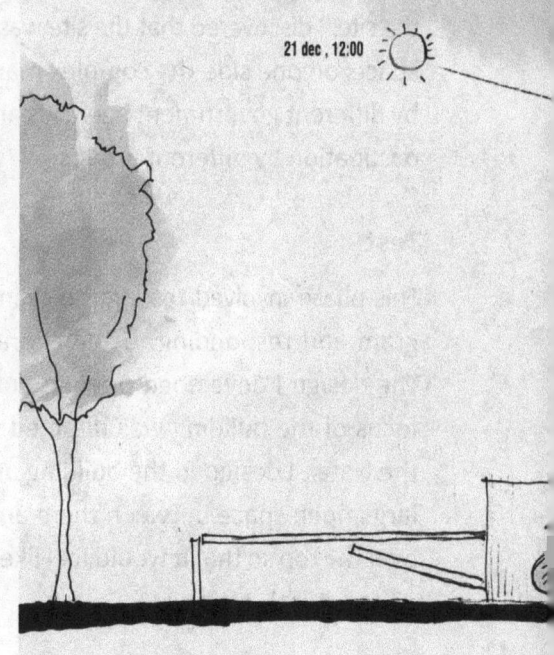

STONEBRIDGE LOCK SKETCH

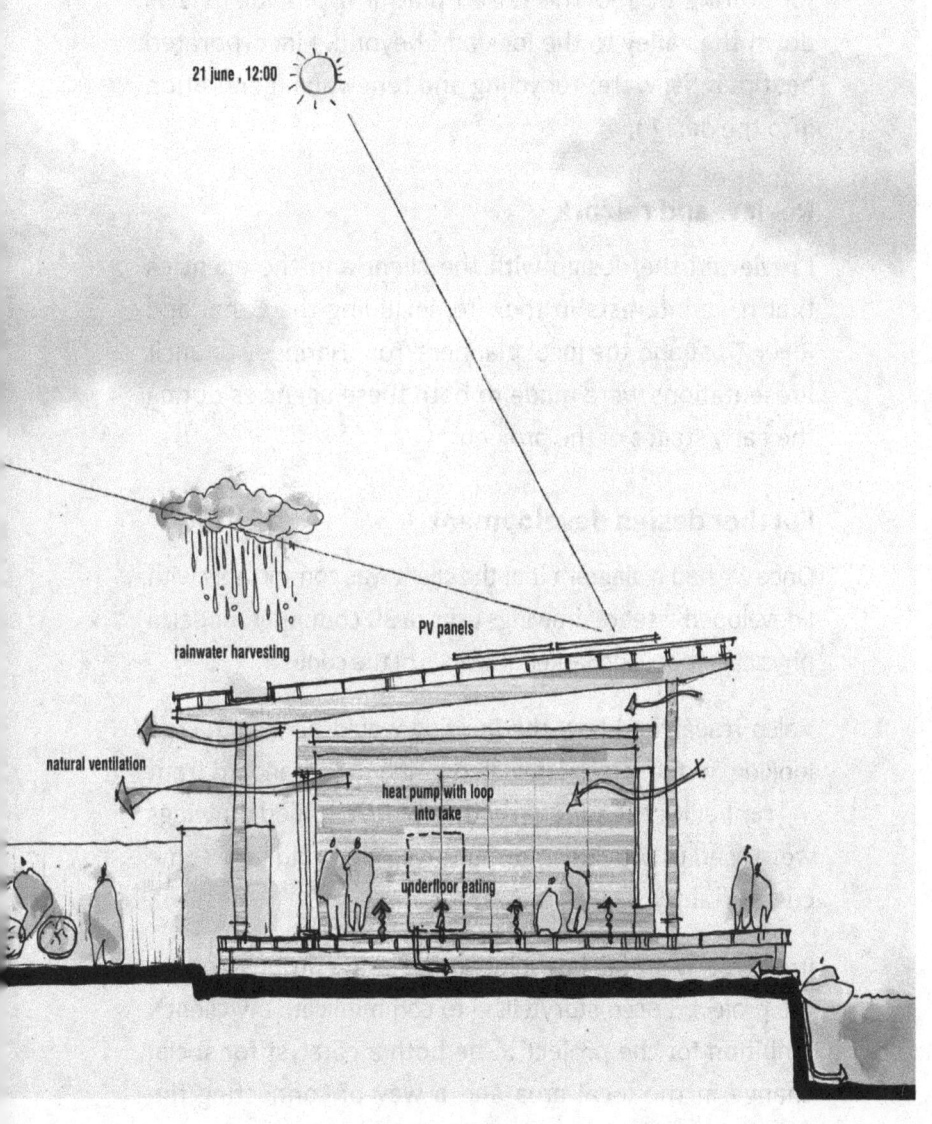

and the impacts of climate change, and a lower platform for storing boats. The raised platform provides views down the valley to the lock and beyond. I incorporated heat pumps, water recycling and renewable generation into the design.

Review and rework

I reviewed the design with the client and the agencies that have interests in the site, including the Canal and River Trust and the local planners from Haringey Council. Presentations were made to both these agencies during the early stages of the project.

Further design development

Once we had a diagram that the client was comfortable with, I developed a set of drawings using a 3D computer model, a physical model and larger drawings of the context.

I also researched how the building would be constructed, looking at both cross-laminated timber and standard framing timber for the superstructure. More detailed drawings were used to provide information so we could gain some cost certainty.

In this early stage, I developed the value proposition for the project. I used storytelling to communicate my client's ambition for the project to be both a catalyst for social change in the local area and a way of connecting the community. He wanted to create a democratic, inclusive space that supported educational programmes that anyone could access and at the same time create employment opportunities for local people.

Another important aspect of the project was developing a design that responded to the position of the sun and local wind patterns, and considered water recycling, sustainably sourced timber and the use of a heat pump energy system that used the canal to improve its efficiency.

To help me navigate the complex planning issues, I put together a small team prior to the pre-application submission. I engaged a planning consultant with specialist knowledge of the local area to help me address all relevant aspects of planning policy. The proposed pavilion was in the green belt, so by its nature it was controversial, and the main challenge was to overcome planning objections.

The council regeneration team and the design officer supported the development, but the planners initially advised we would have to demonstrate that there were special circumstances that outweighed any harm the project might cause, as we were proposing to build in the green belt. My role was to present to the planners what those special circumstances might be and then ensure they became part of the design solution.

Conclusion

I researched the site and conversed with many locals and officials to help me understand the social and historic context. Following this initial phase, I used storytelling to

> communicate my client's vision and to help me co-create a design with the engineers and the local authority. The thematic concerns were related to scale and connections to the larger context. The drawings were tools that supported the proposition and reinforced the small scale of the proposed building in relationship with the immense scale of the local infrastructure, which included pylons, roads, canals and reservoirs. This helped me argue the case for the project.

Summary

One of the secrets of good design is first having a vision for your project and a set of goals that you would like to achieve. This will assist your design team in developing a clear diagram that encapsulates the key project ideas and the most important spatial experiences. The diagram should show how the rooms and circulation spaces relate to each other, how the services work and any important connections, such as roads or pedestrian routes, both inside and outside the building. The diagram must also solve the agendas of all the key players who are interested in the project.

Simplicity takes a lot of creative work and it is likely your architect will work through many options with you before a clear and legible solution is discovered. It will involve test fits, making models, consultation, workshops and brainstorming. These early stages, in particular the feasibility stage when the building diagram is being developed, are where you can have most influence over the design.

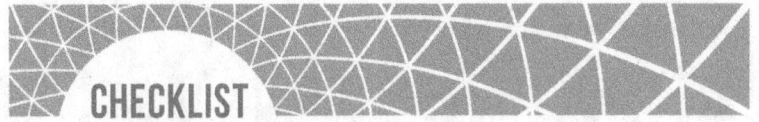

CHECKLIST

- A simple diagram incorporating structural concepts, services and layouts of accommodation is an important milestone in a project.

- Pavilions are a great example of simplicity and this typology of building is used for many sporting clubhouses.

- The clubhouse, gym and changing rooms need to be designed so that access from one to the other is easy and there is a strong relationship with the playing fields and landscape.

- The research phase provides you and your team with all the information they need to understand the opportunities and constraints of your site prior to commencing design work.

- Understanding the context is critical prior to engaging with the local authority.

- Integrate historic and existing buildings into your plans to enrich the experiences you offer.

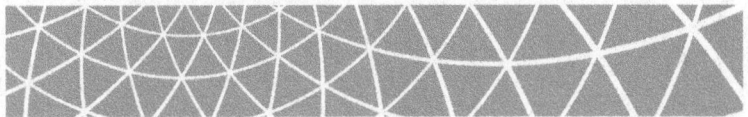

ACTION STEP

If you are thinking about a project associated with your school estate, before you commence the design stage, take our 'How prepared are you to commence an upgrade of your school facilities?' scorecard.

If you are commencing an upgrade of your sports club, before you commence take our 'How prepared are you to commence an upgrade of your club?' scorecard.

You can find this at
www.grahamfordarchitects.com/scorecards

SEVEN
Precision

Precision is critical in the early stages of the design process so the design accommodates your brief. As previously discussed, your brief needs to set out in some detail what you require, what is non-negotiable and what is important technically – acoustics, for example. The drawings that are developed for planning also need to be accurate so you don't have to go back to the planners to change the project because something does not fit when the technical design is developed. What is shown on the planning drawings will be built, and a lot of consultation and co-ordination with other consultants is required to make sure everything has been considered.

Precision is also an important aspect of the research phase of the project. For example, your team should have precise data about where the sun hits your proposed building at different times of the year and how you are going to screen it out in summer or use it in winter. As much data as possible on wind should be gathered to inform how the building will respond to wind conditions. It is also important to have information on plants, wildlife, landscape, water and hydrology.

Precision is important in the preconstruction stage when your team produce technical documents the builder will use on the construction site. These documents embody craft; they describe how materials meet, and how the finished building will look and feel.

When I studied architecture design, discussions were centred around typology (types of building) and spatial issues, but there was less emphasis on materials, and building technologies were taught separately to design. Materiality (the materials the building is made of) is not just about construction but is also about the very essence of the building, what we would call tectonics or the art of construction. Tectonics encompasses the structure, the external skin and how precisely materials are assembled and fixed. Tectonics are fundamental to determining the form (shape), spatial complexity and sustainability of your project.

You should choose an architect who is interested in both the aesthetic and formal resolution of buildings, as well as craft. They will undertake research with product manufacturers and subject matter experts during the design phase. In the later phases, you will have important discussions with the manufacturers who make various parts of your project.

Tacit and explicit knowledge

As we saw in the discussion on spatial intelligence, an architect draws on their tacit and explicit knowledge. Their training is a type of apprenticeship involving hours of meticulous work in the studios of well-respected architects who have commissions to build. This is where the founder's individuality and tacit knowledge dominate. The studio glues people together through work rituals and mentoring. It is in this space that architects invite collaborators, contractors and clients to join them. These people bring their explicit knowledge to the process and knowledge transfer occurs, building up a shared expertise about how to approach and put together buildings of some complexity. This knowledge is then tested and refined during construction.

The architect's job is to co-ordinate all the information for the project. This is a significant skill that is frequently overlooked by both clients and contractors. In many projects there are flows of information across continents, and parts of buildings are manufactured all over the world. The co-ordinator understands the roles of all the other participants and integrates their designs. The architect's general arrangement drawings are like maps that are used by consultants, contractors and clients to navigate the project.

As architects work on new projects, their memory bank of images and their ability to imagine space gets updated. This tacit knowledge is a resource that your architect uses to help design your building. If your architect has knowledge of craft, they will understand the spatial implications of details, the hierarchy of details and how to join different materials. All these elements are combined to make your project special. Architects imagine how materials will meet in 3D space, what that detail will look like and how the contractor will build it. It is important to engage craftspeople in the process of hand sketching, discussion, debate and refinement of details. In the next case study, I will discuss my involvement in the construction of a pavilion on the Olympic Park.

CASE STUDY: THE BMW PAVILION

In 2010, I was employed to help Nussli, a contractor who specialises in the construction of buildings in the sport and leisure sectors, to work on the BMW Pavilion. The contractor needed a local architect to be a link between the consultants and the subcontractors, and to co-ordinate the design on-site. The pavilion was located inside the Olympic Park and was built over a canal.

I was employed for my experience of English procurement, knowledge of available products and my network of specialist subcontractors. I also had knowledge of the Olympic Park and how it was organised. I was involved in procurement, interviewing subcontractors and liaising with the English consultant engineering team that was based in the UK.

The first challenge on the project was that the client and the architect were based in Germany and visited the contractor's design office in Stratford once a week. The contractor was Swiss and when I joined the project they had just arrived in the UK.

My role was to interpret the executive architect's design by meeting with the project architect regularly over many weeks. I was able to collaborate effectively with him as he was confident I understood what BMW wanted to achieve. Once I was familiar with the drawings, I communicated the design intent to the contractor and the subcontractors.

The design included complex systems that needed to be co-ordinated, including a water roof, a waterfall down the façade and water purification processes. The pavilion also needed air conditioning in glass pods for prototype cars, specialist timber work for the curved canopies, and a full internal fit-out.

I assisted with the selection of subcontractors for key areas, such as the steel and timber structure and the curtain walling. The other big challenge was that, to meet a tight programme, the contractor was proceeding at pace with information that was still in development. I communicated the design intent of a complex bespoke pavilion to a supply chain that had no established relationship with the main contractor.

I co-ordinated my small team of consultants, which included the mechanical and electrical design manager and a local architectural firm who were assisting me with the analysis of the project, unpacking the design and drawing details as required. My team also included an expert in hydrology, who was advising me on issues of airborne bacteria and public health relating to the water roof and waterfall over the façade to ensure the installation was safe.

To facilitate this co-ordination and to communicate the design, I managed the specialist subcontractors' design packages. I attended co-ordination workshops where I shared sketches to help clarify details and interfaces with other trades, and I reviewed their shop drawings.

> These workshops involved the exchange of subcontractors' explicit knowledge of their specialist area for my detailed knowledge of the design intent of the project. I was a bridge that communicated concepts from the design office to the building site.
>
> As a design advocate, I made judgements about how best to protect the design intent and at the same time I needed to be practical and agree to changes that did not impact the design to ensure the project was delivered on time.

Accurate costing

Accurate drawings are needed so your project can be costed. As we develop the design, we will produce more detailed drawings of the spaces and we will describe the quality of those spaces, the materials, the look and feel. Together with your design team you will select the technology that supports your hardware, based on what user experience you want to provide for your customers. The quantity surveyor can only provide the cost information you need if the design is fully developed and accurately drawn. This is another reason why precision is so important.

It is critical that you use the quantity surveyor's expertise throughout the project and in this way, the project will remain under control. The cost information that is collected during the project phases will be used to establish a contract with the builder and assist with the administration of the project during construction.

Summary

Precision is critical in all phases of the project, commencing with your brief. In the preconstruction phase, accuracy is built into the planning drawings and then into precise construction documents. This process involves knowledge of construction and design combined with knowledge of your brief, the local site and planning rules.

The transfer between tacit and explicit knowledge and the unique conditions of your project and organisation is where new knowledge is created. It is also where craft is embedded into the project. The different roles I have adopted have helped me collaborate with fabricators, as demonstrated by the BMW Pavilion.

Regardless of the procurement route, construction drawings must be fully detailed, precise and co-ordinated with the engineering design, and they should anticipate the logic and sequencing of construction. Poor documentation leads to a poor-quality product and delays, as decisions will be made as construction goes along. The benefit to you is that having a well-researched set of construction documents ensures a smooth-running process that saves time and money.

CHECKLIST

○ Precision is the second principle that is critical for the success of your project.

○ Precision comes from careful crafting of the drawings from the beginning of the project.

○ Precision helps to protect the design as it ensures the contractor has sufficient information to understand how all the parts come together.

○ Precision prevents delays and problems on the construction site, making the build much smoother.

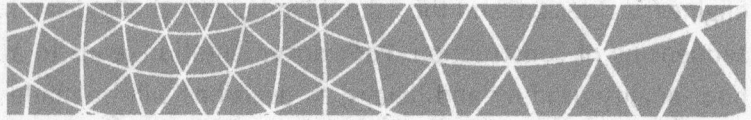

EIGHT
Efficiency

All projects need to be well organised. At the beginning, consider how you are going to procure the building with your architect, who your contractor might be and what information you will produce before you enter into a contract with them. You may want to involve a contractor in discussions early on so you can use their expertise in deciding what systems to build with. Once planning is gained, you will move into the preconstruction phase where certain aspects of the build can be accelerated so the project can get underway early.

If your team is advising on integrating manufactured components into the design, this can save both time and money but it needs to be planned from an early stage. One final thing you need to consider with your design team is how technology will help you to construct and manage your facility more efficiently. This will include BIM and the internet of things (IoT), encompassing network-connected devices such as sensors, as well as network data analytics of lighting, indoor air quality, noise, temperature and energy use.

If parts of your project can be manufactured using some of the technology described above, you will benefit from products that have been improved over time in factory conditions. These could be timber panels, wall systems or structural platforms, where embodied carbon has been reduced through improvements to the design. Efficiency is part of the tectonic nature of the building and will determine its form (shape) and character. You may find prefabrication also improves efficiency on-site, which reduces build time and energy consumption. Sustainability is about the ability to adapt, change and relocate buildings, and prefabrication assists this process.

Our nine-step process takes you from feasibility to completion and is based on the work stages outlined by the Royal Institute of British Architects (RIBA). These steps include:

- Discovery (RIBA Stage 0)
- Feasibility (RIBA Stages 1)
- Concepts (RIBA Stage 2)
- Final proposals (RIBA Stage 3)
- Preconstruction (RIBA Stage 4)
- Negotiation and contract (RIBA Stage 4)
- Construction and delivery (RIBA Stage 5)
- Handover and post-occupancy evaluation (RIBA Stages 6 and 7)

These logical steps make your journey efficient by creating a map that is easy to follow. By knowing what happens at each phase, you will feel in control and everyone will be clear what the deliverables are at each stage.

Designing a project and having it constructed can be a complex and difficult process unless you have done this many times before. There can be no confusion on a construction project; you will want to know the headline issues of scope, specification, time and money at each step on the process map. The project must pass through all of these steps logically and the milestones be met before advancing to the next stage.

You need to be clear about your priorities. If speed is essential and will make a big difference to your business operation, then efficiency might be the most critical principle guiding how the project is managed. It is possible to sequence the preconstruction design work and accelerate the structural design if you have a well-co-ordinated and precise set of drawings from the final proposals (planning) stage. This will allow the project to get underway as soon as possible.

You should be aware of two considerations that may prevent you from commencing on-site. The first is the planning conditions your team may need to discharge before the project can commence. The second is the Party Wall etc Act.[9] Both these processes take time and need to be factored into your overall design and construction programme.

Precision, efficiency and sustainability

One of the aims of design for MMC is to use mass production to improve efficiency, delivering custom-made products at low cost. New ways of procuring buildings are now possible if your project manager works more directly with manufacturers and purchases at a component level to speed up the process and reduce transactional costs.

A more precise, design-to-manufacturing approach helps to reduce the amount of material used in each component. It is also more efficient to reduce the time spent on-site and the amount of waste generated. One of your considerations when

---

9   The National Archives, 'Party Wall etc. Act 1996' (1996), www.legislation.gov.uk/ukpga/1996/40/contents, accessed 5 November 2022

114   The Total Environment Masterplan

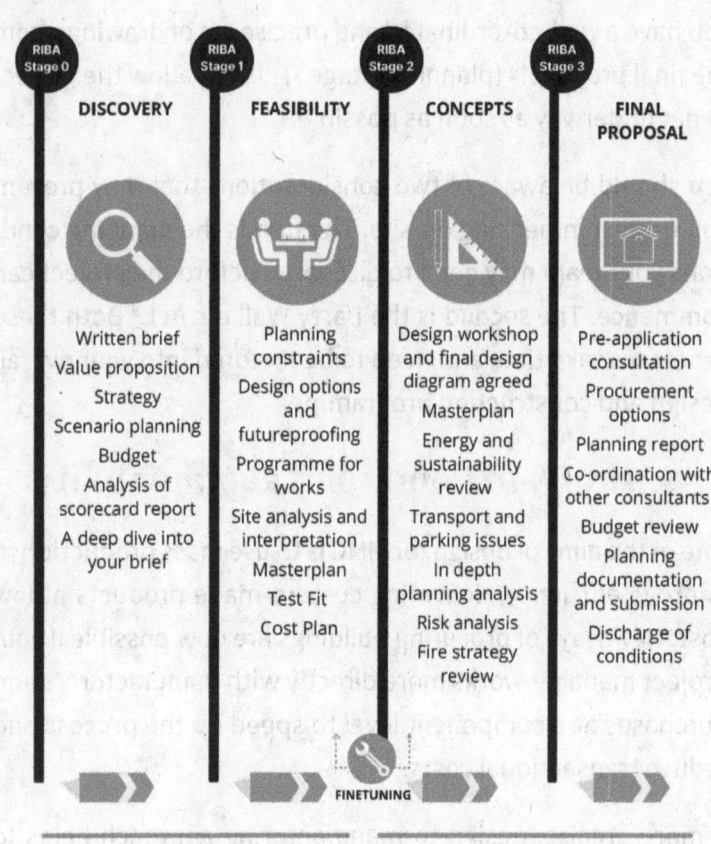

Efficiency 115

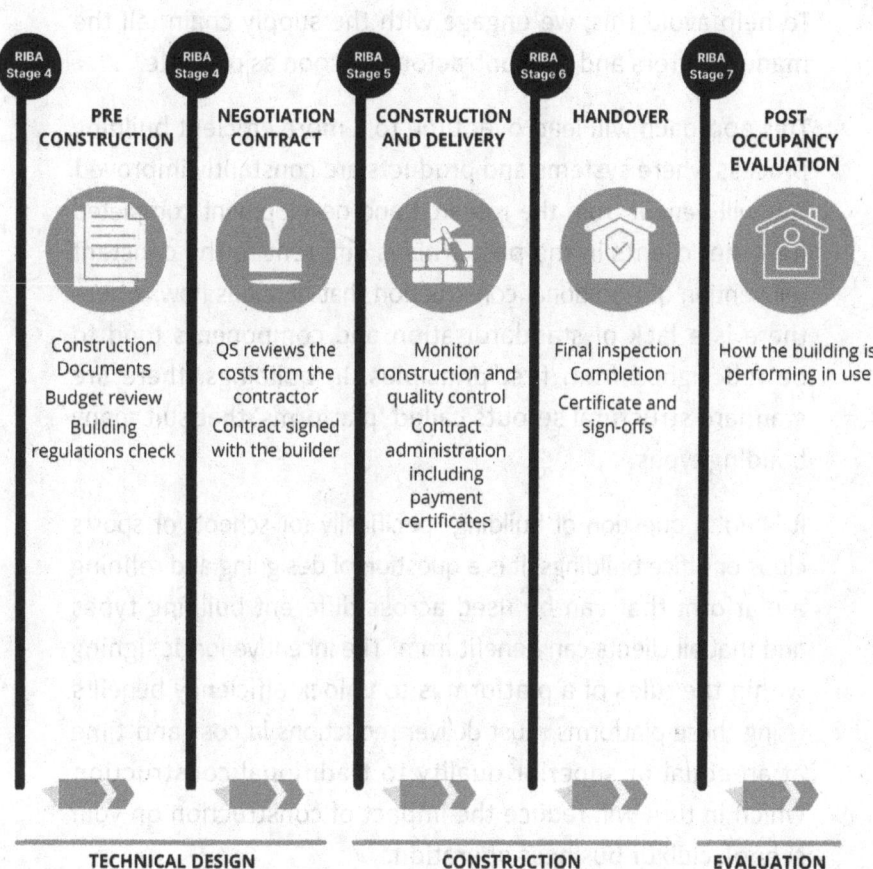

RIBA Stage 4	RIBA Stage 4	RIBA Stage 5	RIBA Stage 6	RIBA Stage 7
PRE CONSTRUCTION	**NEGOTIATION CONTRACT**	**CONSTRUCTION AND DELIVERY**	**HANDOVER**	**POST OCCUPANCY EVALUATION**
Construction Documents	QS reviews the costs from the contractor	Monitor construction and quality control	Final inspection Completion	How the building is performing in use
Budget review	Contract signed with the builder	Contract administration including payment certificates	Certificate and sign-offs	
Building regulations check				

TECHNICAL DESIGN — **CONSTRUCTION** — **EVALUATION**

OUR 9 STEP PROCESS

designing a building should be what will happen to it once it is has served its practical life. The fact that design development traditionally took place in isolation from the supply chain is a significant source of missed opportunities to optimise the design and leverage top-class construction knowledge. To help avoid this, we engage with the supply chain (all the manufacturers and subcontractors) as soon as possible.

This approach will lead over time to a more efficient building process where systems and products are constantly improved. You will benefit from the research and development completed for other clients in the past. This is different to the constant reinvention of traditional construction that happens now, where there is a lack of standardisation and components tend to be redesigned from first principles. In buildings, there are standard structural setouts called 'platforms' that suit many building types.

It is not a question of building specifically for schools or sports clubs or office buildings, it is a question of designing and refining a platform that can be used across different building types and that all clients can benefit from. The incentive for designing within the rules of a platform is to unlock efficiency benefits. Using these platforms must deliver reductions in cost and time at an equal or superior quality to traditional construction, which in turn will reduce the impact of construction on your school, club or business operation.

Prototypes and optimisation

On average, office buildings in the UK go through forty different internal changes throughout their life. These internal elements (partitions, doors, ceilings etc) have embodied energy, so the sustainability story needs to articulate how, during the operational phase of the building, they can be reused or recycled.

One of our core ideas about sustainability is the design of buildings as products that can be adapted, extended, improved and even relocated over time. This applies to both the external envelope and structure and to the internal partitions and spaces. These ideas need to be embedded enough in design thinking for your building to be optimised and improved to suit your and your customers' needs over time.

To optimise your building during the operations phase of the project and improve efficiency, it should be possible to adapt the internal space without having to demolish walls or ceilings or compromise the acoustics and mechanical and electrical installations. This approach will ensure high levels of satisfaction from the space. Optimisation involves your team designing a tech-enabled smart asset that includes:

- The hardware (the building)
- The software (digital tenant / member experience platforms)
- The services (hospitality, leisure facilities, meeting rooms, medical facilities etc)

The more efficient your build and the easier it is to adapt an existing building, the less impact the construction will have on your team, pupils, tenants or members.

▶▶ CASE STUDY: THE HYDE PARK BOATING PAVILION ◀◀

My client had been running his boating operation in Hyde Park for several years and wanted to upgrade his facility to make it more efficient, provide better accommodation for his staff, upgrade the jetties and create a small retail space.

My role was to develop the design, apply for planning amendments from the local authority and then develop the technical drawings and details for construction. Before starting design, I researched the project by interviewing fabricators who had manufactured timber panels and steel structures for similar types of pavilions. I interrogated their shop drawings to gain inside knowledge of how their products had been put together, including the details and exact dimensions of panels and frames.

During design development, I worked with a team of engineering experts to help me deliver the design. I introduced sustainable cladding and insulation products and a 'kit of parts' building system. The building design went through further development with a fabricator, who manufactured factory-produced timber panelling, windows and doors. The design information produced through this collaborative process enabled me to integrate craft and precision into the building.

The pavilion was conceived as a kit of parts so that it could be constructed off-site and easily modified, and be flexible in the way the interior was arranged. The demountability of the pavilion is a unique solution I developed together with the structural engineer in response to my client's brief that requested the building be easily removed from the site at the end of his twenty-year lease.

HYDE PARK BOATING PAVILION

Building information modelling

The future of construction will revolve around the creation of new data and a technology-driven process which includes automation. Adopting this technology will increase efficiency during design and construction and will assist the management of the building.

Design teams are now using BIM during design development. The challenge going forward is to fully integrate the mechanical and electrical design into the BIM model. This process is developing and becoming more mainstream for larger, more complex projects. Before you decide whether you should use BIM, consider carefully the benefits for the size and difficulty of your project. There are considerable costs associated with BIM and you need to be sure that the advantages are clear both now and in the future. A fully integrated Stage 3 BIM model may well not be suitable for smaller projects.

For larger complex projects, 'soft landings' – linked to BIM – bring asset, property and facilities managers together with the world of construction. A soft landing means that when new buildings are handed over, all the information required for a smooth operation is available. Your team will need handover training conducted by the contractor's team, who should be contracted to be available to assist your facility managers for the first few weeks of operation. Facility managers are now more engaged in the briefing process from an operational and maintenance point of view, and they need data they can use to operate the building. In this way, they will be better able to assess building performance.

In the future, buildings will have sensors and be managed by building technologists with digital models that will give them real-time feedback on the building's performance and the exact location of any malfunction. You should think of your building not as walls and a roof but as many systems interacting with each other, such as windows 'talking' to the lighting and the air conditioning systems to adjust output and reduce energy consumption. This IoT management will make buildings easier to maintain, repair, replace and relocate. There will be better control and data for feedback and analysis. You should ensure that energy monitoring is installed to provide the data you need so your heating and cooling systems can be optimised to reduce energy consumption.

Information release and design changes

Efficiency in construction is about having the information available to the contractor when they need it to ensure the project keeps moving. The biggest problems I have seen on construction sites were caused by a lack of quality design information or clients changing their minds and changing the design during the construction process.

Information needs to be carefully managed so you don't get overwhelmed by having to make too many decisions at the same time. If you are building a basement under a building, at the start of the process you only need information on how the levels are set out, drainage, structure and room layouts. Once the basement dig is underway, further details for the fit-out of the interior can be developed, including lighting design and finishes. On all projects, it is important to know what information is required and when so you know what decisions you need to make and when you need to make them by.

Commissioning and post-occupancy evaluation

Commissioning is where the mechanical and electrical systems, including boilers, heating and ventilation, and lighting controls, are tested and optimised to make sure they are working correctly. The critical final step in efficiency is ensuring all your systems work before you occupy the building and that you have good-quality information to help you manage the new facility post-completion. At the end of the job, you will receive a completion certificate from building control. The final items that are important are the health and safety file and the operation and maintenance (O&M) manuals that have detailed information about how your building's systems (plant, boilers etc) operate, warranties and test certificates.

You should always obtain a set of as-built computer-aided design (CAD) drawings, and the BIM model if one was produced by the design team, to ensure you have accurate information about your asset. This data will be critical for maintenance and for any adaptations and changes you may want to make in the future.

The final step involves a critical evaluation of the building in use. Ideally, the design team returns after twelve to eighteen months to discover what works and what does not work. Post-occupancy evaluation (POE) is a way for you and your designers to assess how the building works in use. This will guide any enhancements that might need to be made and inform your next project. POE helps to articulate the intangible but centrally important social and organisational value that the design team delivers to you and your clients. For example, how does the building design encourage social interaction, promote health and wellbeing and reduce energy consumption? A POE will identify strengths and weaknesses in the design. This includes:

- Accident rates in the facility
- Energy efficiency

- Data about internal environmental quality
- Information on health and productivity gathered through surveys

It is an important way to promote better co-operation with facilities managers who are already collecting and assessing operational information.

For low-energy buildings, you will need to collect data from electrical energy monitoring, which includes a limited set of other parameters, such as temperature, CO_2 levels and humidity. POE enables you to understand how your building is performing so you can compare this to your design brief.

Summary

Efficiency is a critical principle as it reduces the impact of construction on your operation. Every week a contractor is on-site is a cost and potential disruption. This is why good planning, anticipating problems in advance and using prefabricated materials and off-site manufacture whenever possible can result in improved efficiency during construction.

These prefabricated systems are not appropriate for all projects. They were chosen for the Hyde Park Boating Pavilion to speed up the construction process, reduce the impact on the park, and facilitate removal at the end of the client's lease.

It is important to adopt technology wherever possible as doing so provides you with valuable data that can be used for both maintenance and operation of the building and for any changes you would like to make in the future.

A key conversation with your design team will be around 'long life and loose fit', which means making sure the building is adaptable for future internal configurations. Innovation and quality start with the designer and flow from them to other members of the team.

POE rarely happens with buildings. Other consumer products, such as cars, are continuously improved when each new model is produced. The problem with buildings is that they are currently all prototypes and only a few designers and clients are learning the lessons from the last model produced. The future of the industry is in making more repeatable, standard products that can be continuously improved over time. In this way, we can all learn lessons from the end users, feed this information back and improve the next model.

Efficiency is about adopting new technology, being innovative, being organised, producing quality design information and working as a team.

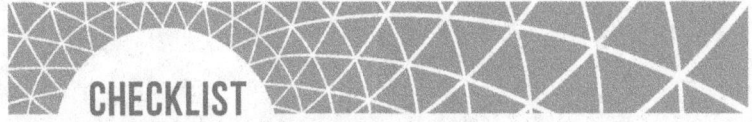

CHECKLIST

○ A logical nine-step process should be followed when designing and constructing a building.

○ Efficiency is about planning everything in advance so all the information required to build is available at the right time.

○ Scope, specification, time and money are the headline issues at each stage.

○ Planning conditions and party wall matters are often overlooked and can considerably slow down your project if they are not considered early.

○ Prefabrication will make your project more efficient, reduce waste on-site and reduce the impact of construction on your school, club or business.

○ Use the IoT where possible to help you reduce energy consumption and improve the wellbeing of staff and visitors.

○ Take advice about whether BIM is suitable for your project. Will BIM make your operation more efficient? Carefully assess the costs and benefits before proceeding.

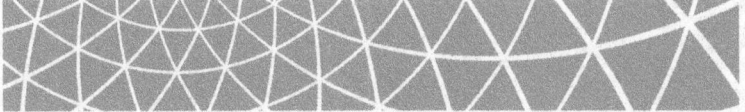

NINE
Collaboration

Our fourth principle, collaboration, involves building a great team of engineering consultants, contractors and your advisors and staff. A well-functioning team will work together to overcome the challenges you may face with planning and construction.

Collaboration and building partnerships with quality consultants and suppliers is essential for masterplan projects and sustainability. These larger scale projects require a lot of 'joined up' thinking that includes infrastructure, transport, landscape, hydrology, biodiversity and energy generation among other disciplines.

Your architect is like the conductor of the orchestra who does not need to play every instrument but does need to know the qualities of each one and how to combine them to create a compelling performance. One of your architect's greatest values is their ability to understand and synthesise diverse and complex inputs across many disciplines, and forge these into one compelling experience. Like the conductor, it is necessary to understand the unique skills of all the different consultants and combine these talents to create a total environment. As we discussed in Chapter Two, before submitting a planning application your team should introduce your ideas and proposals to the local planning authority. You will want your design team to engage with the council officers responsible for highways, heritage and all other aspects of the application. You should consider the planners and their specialist support teams to be consultants to your project team, helping to make sure the best possible solution is found.

Even before planning permission has been gained, you will want to collaborate with contractors and manufacturers to facilitate brainstorming and innovative thinking and resolve issues of logistics and buildability.

Roles to facilitate collaboration

I have already introduced you to the different roles I adopt on projects. Performing these roles involves engaging with government agencies responsible for the site – for example, the Royal Parks and the Canal and River Trust; it also involves collaborating with planners, regional parks and local people to help understand their concerns and the impact development might have on their area. Collaboration requires radical empathy, active listening, engagement with subject matter experts and being open to different points of view.

Your architect will lead the team during the precontract phases of the project and, using a form of storytelling, translate your strategic brief into a physical vision for the building and landscape. In these early design stages, your ideas and vision, architectural considerations, engineering, logistics and manufacturing know-how will be integrated into the building's design.

Collaboration with the contractor

Maximum innovation occurs at the early stages of a project. According to Jason Millett,[10] CEO of Mace Consultancy, 'in the words of one senior infrastructure executive, "90% of savings and innovations actually come within the first 10% of project spend."' With projects often focused on progressing to the next stage as quickly as possible to secure more funding and 'early wins', many innovations are overlooked at the beginning and then prevented from being introduced at a later date by the constraints imposed in earlier stages and the resultant planning process.

It is by tapping into the contractor's knowledge that we can make important decisions about constructability, logistics, possible structural systems, MMC and other pragmatic considerations. Having a sensible approach to procurement and drawing up a comprehensive contract is the best way to achieve your goals in a collaborative process.

Many clients are concerned about paying too much for construction. To alleviate this fear, you should work closely with your quantity surveyor and benefit from their experience of building in your sector. They will correctly measure and cost the project, so you can be confident that the budget is realistic. Almost all construction costs are well known and documented. The next case study illustrates how collaboration helped in the construction of the Handball Arena for the London 2012 Olympic Games.

---

10 J Millett and D Dabasia, *A Blueprint for Modern Infrastructure Delivery: Why things go wrong and how to put them right* (Mace Insights, 2020), www.macegroup.com/perspectives/201112-insights-blueprint-for-modern-infrastructure, accessed 30 October 2022

▶▶ **CASE STUDY: THE HANDBALL / COPPER BOX ARENA FOR LONDON 2012** ◀◀

The building projects for the Games were procured with a design and build model using an NEC3 contract[11] that prioritised collaboration between the contractor and the design team.

The technical design was developed by the contractor's design team, which included their executive architect Populous, a specialist sports architect who also designed the main Olympic Stadium. I was employed by Holmes Miller Architects to assist with quality assurance. This role involved the co-ordination of a team that included the design architect, the mechanical, electrical and structural supervisors, and the project's client.

The executive architect's scope was to produce a full package of drawings that could be used for construction purposes and to help the subcontractors develop their contractor design packages. My role was to interpret the design intent and detailed specification prepared by the client's concept architect, and assess this information against the construction drawings, the shop drawings produced by the subcontractors and what was being constructed on-site. I was the link between the client, the concept design team and the contractor's team.

---

11 NEC, www.neccontract.com/products/contracts, accessed 31 October 2022

This collaboration was facilitated by having the entire team housed on-site in accommodation a short walk from the construction site.

The project management team were appointed to ensure that the building was delivered on time and on budget. I would regularly inspect the site with the contractor's quality team, the client and the project manager, and we would agree whether the quality of the subcontractor's work was acceptable or not. Before a new trade commenced, I would review a sample of the work and agree on any changes that had to be made. This initial element in the building process became a benchmark for the quality required for all future work. For example, site visits on this project also involved reviewing the fabrication of the copper wall cladding. I inspected the installation of the copper sheets and witnessed the subcontractor's quality assurance procedures. Through interpreting the technical drawings and visiting the site every day for eighteen months, I became familiar with the requirements for the design of stadia.

Summary

A design project involves collaboration right from the start of the process. You will collaborate with the design team, who will also then collaborate with the local planning authority.

To end up with a remarkable building, you need to appoint a team that can develop innovative solutions with you. This team will include engineers led by the architect who has an overview of the entire process.

Construction is complex and the risks need to be managed, mitigated and tracked as the project progresses. Quality information should be provided when needed to make sure the contractor can keep the project moving and the final product is as good as it can possibly be. Intuitive leaps happen when subject matter experts, clients, fabricators and designers whose knowledge can be harnessed to the benefit of the project collaborate in a shared space with equal status.

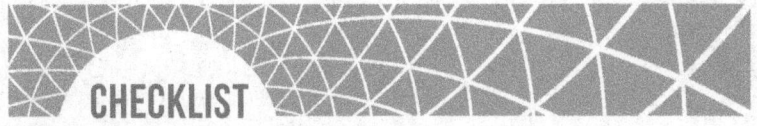

CHECKLIST

- ○ Collaboration happens at all stages of a project, starting from the writing of the brief.

- ○ Talk to the local planning authority as early as possible to discuss your project.

- ○ Contractor knowledge is a powerful asset you should tap into as early as possible.

- ○ Construction is a team game and innovation solutions come from collaborative workshops where all participants are treated as equals.

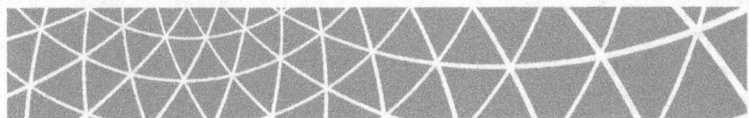

TEN
Sustainability

As an industry, the property sector is responsible for a third of global greenhouse emissions, and we consume 40% of the world's energy.[12] Given the compelling evidence of climate change, we now have the ability to provide better solutions and build exemplary projects that others can follow. If your asset is not sustainable and does not run with modern energy-efficient systems, its value may rapidly diminish and you may not meet the proposed statutory requirements for energy performance.

You will want to know how your building gains a good energy performance rating and how, in the near future, it can become zero-carbon. As discussed in previous chapters our aim is to achieve this through the tectonic nature of your building and using materials that can be reused and replaced easily to assist with maintenance. You should use local materials where possible to minimise costs of transportation; you should bolt your building together and screw in boards rather than nail them so all the parts can be easily taken apart and reused. This approach will help reduce the energy you use and therefore your carbon emissions. Increasingly, your customers will choose which venue they use or visit based on your building's sustainability credentials.

---

12 UN Environment Programme, *Sustainable real estate investment* (UN Environment Programme, 2016), www.unepfi.org/industries/investment/property-publications/sustainable-real-estate-investment-2, accessed 5 November 2022

We are all facing complex challenges as a result of the human population growing rapidly and putting huge pressures on ecosystems. We need to accelerate the way we sustainably produce buildings to keep pace with the demand, and at the same time develop clean electron production to reduce our dependence on fossil fuel energy and reduce carbon emissions into the atmosphere. It is critical to consider how MMC can be used to reduce waste, improve flexibility and help make your building more efficient when assessing how sustainable your project will be.

One of our missions is to design buildings with reduced operational costs. It is your architect's job to ensure building resilience against extreme weather conditions, such as heatwaves and floods, and critical events, such as earthquakes. A strong proposition that incorporates sustainability will help convince the local planning authority that what you are proposing will make a positive impact on the area.

The Todd Corporation, one of New Zealand's largest and most successful companies, has interests in hydrocarbon exploration and production, electricity generation, energy retailing, property development, minerals, healthcare and technology. In 2000 I was granted a Todd Foundation Scholarship, and the fieldwork I completed in this trip to the US and Europe was the basis for my master's research. The funding allowed me to travel the world and interview leading engineers and architects who had completed exemplary sustainability projects. Many of the architects I visited had pioneered a close relationship with their engineers, and projects were developed by introducing the first principles of environmental design at the concept stage.

In 2002, I returned to the UK and was employed by Wilkinson Eyre to help with the design of the Jodrell Laboratory, a new research and office building in Kew Gardens, which won a Royal Institute of Architects award in 2008. This project allowed me to put into practice many of the lessons I had learned, including how to introduce thermal mass into the building design to reduce heating loads and how to design large buildings with mixed-mode ventilation.

A visionary masterplan

Anyone who manages a large estate of buildings will be thinking about development over the next ten to twenty years and how best to allocate their precious resources to improve their assets. If you are a bursar or head teacher you will have:

- A vision for the teaching and learning pedagogy to which your school is committed
- A strategic plan
- A physical masterplan

With your design advisor or architect, you will want to link your physical masterplan to the school's strategy and its learning vision. These three documents work together and involve cycles of analysis, goal-setting, fundraising and actions. Each informs the other so that the planned facilities support your specific approach to learning but are adaptable enough to be aligned with evolving approaches to pedagogy.

If you own a sports club, your design team will need to know your current membership numbers and who joins; any communication you have had with the county council, local authority or Sport England; what funding you have available; if any market research has been undertaken; and your timescales. Prior to design work, research needs to be carried out to assess what facilities you currently offer, what the community needs, what your members want and what can help generate more income and future-proof your facility.

A masterplan is important so you don't construct a building that solves a short-term problem but creates problems for you in the future by blocking light or access to parts of your site. You need to be sure you don't locate a building somewhere that might be better designated for other uses. You will need to consider infrastructure requirements for future developments, including drainage, electrical cables and IT.

You should discuss who you need on your team to produce the masterplan with your design advisor or project manager. If you are a head teacher, you may already have an in-house team that consults with school staff, the board of governors, students and the broader school community, including parents. The consultation phase is vital to help inform and develop the brief. A good brief, as we saw in Chapter One, is the basis of all future design work and will result in a successful masterplan that reflects insights and perspectives on education and design. All ideas should be evaluated against your school's vision for teaching and learning, and your strategic plan.

This plan must be visionary and at the same time have a strategy for which projects to develop in what order. You will also be discussing how individual projects would be funded.

Similar to the design of a building, in a masterplan you must think of key adjacencies between buildings and key routes through the site. The masterplan will look at how your estate of buildings interfaces with surrounding buildings, landscape and infrastructure and it will consider local planning documents.

Like all projects, your value proposition for the masterplan must consider social, economic and environmental values. The masterplan will enable you to communicate your vision to the wider community, discuss your funding priorities and respond when funds become available so you can design new facilities in keeping with your values and beliefs. For example, the masterplan for a school should be created in tandem with your teaching and learning pedagogy and your strategic plan.

Future-proofing your facility

The global pandemic of 2020/2021 showed us that we need to be much smarter and more sustainable about what we do. We have the knowledge and skills to design better buildings and landscapes than ever before. Our mission is to use design to drive innovation and real change and to tackle the degradation of ecosystems. To achieve this, we must be advocates for enhancing natural systems by working more collaboratively with clients, consultants and contractors to find construction solutions that minimise disturbance.

You will probably be thinking about the impact the pandemic has had on your employees and clients and reconsidering the future of your organisation. In a competitive market, now is the right time to reassess your buildings and consider if they are enhancing the wellbeing and productivity of your staff and customers. Do they reflect your brand values and what you stand for? The pandemic was an existential health and safety crisis. Your customers will now realise that your building has the potential to do them a lot of harm. They may want you to upgrade the ventilation systems of your existing building and improve its performance through better data collection. Most people will want to visit establishments that will keep them safe and enhance their wellbeing. Better natural light and better-quality air combined with the improved sustainability performance of your building will future-proof your operation. Customers have options and to attract them you will need to become an accomplished storyteller. Great brands represent great stories and if you are renewing or building new facilities, the story you tell needs to capture the wider benefits of what you are proposing.

Larry Fink, the CEO of BlackRock with assets worth 6 trillion US dollars under management, wrote in his recent annual letter to shareholders that there is a fundamental reshaping of finance occurring and that climate change is driving a profound reassessment of risk, which he anticipated would result in a significant reallocation of capital: 'Investors are increasingly reckoning with these questions and recognizing that climate risk is investment risk.'[13] When the man at the helm of a massive investment fund like BlackRock believes that sustainability and climate-integrated portfolios will provide better risk-adjusted returns to investors, we know that a significant paradigm shift has occurred.

---

13   LA Fink, 'A fundamental reshaping of finance' (BlackRock, 2020), www.blackrock.com/us/individual/larry-fink-ceo-letter, accessed 30 October 2022

You will want to consider what the impact of changes in energy supply and the risks of climate change will be for you and your staff, investors and clients in the future. I predict that clients who truly embrace sustainability will build powerful brands, attract the best customers and generate larger profits.

Future energy requirements need to be thought about during the initial design phase, including how new and potential legislation may impact the energy systems you can use and the allowable emissions. What you don't want is to design a facility that runs on a source of energy that is being phased out and has unacceptable levels of CO_2 emissions. You should consider heat pump technology and, if possible, use renewable energy generated on-site. It is worth remembering that it is much harder to integrate energy systems as a retrofit once the building has been constructed.

Buildings that consume a lot of energy cost a lot of money to run and produce carbon that contributes to global warming. Where possible, ask your architects and engineers to make sure the design will reduce the overall running costs of your project. A key consideration in reducing energy use will be deciding which heating and cooling system will be installed. In the Hyde Park Boating Pavilion project, we used the Serpentine Lake as a heat exchange source for the building. Because water temperature is more stable than air, the heat pump system uses the lake as a heat sink or a heat source, depending on the season. The thermal energy is collected via pipes that sit in the lake and the amount of heat transferred into the building is four times larger than the electrical energy consumed. The heat in the boathouse was distributed using an underfloor heating system. If you don't have a lake you can use, then you should consider an air source or ground source heat pump.

▶▶ CASE STUDY: THE WOOLWICH SHOOTING RANGE ◀◀

During the preparation for the London 2012 Olympic Games, I was employed by Holmes Miller Architects to work on the spectator enclosure for the shooting venue as a quality assurance advisor to the Olympic Delivery Authority. My job was to make sure the venue was being constructed according to the employer's requirements. This venue consisted of three pavilions which formed a collection of colourful tents scattered around Woolwich Common to create a shooting and archery campus. The structural frame was built up from standardised steel trusses that are widely available for hire. These were wrapped with a fabric skin. The range (field of play) was enclosed in a plywood perimeter fixed to a steel frame. Most parts of the pavilions could be easily removed and reused. The piles for the venue were all made from recycled gas pipes driven into the ground – easy to recycle once the Games were over. The structure for the three pavilions used rental components. This approach is called precycling, where you plan the second and third uses of the equipment during the initial design phase. This innovative thinking can be more widely applied for different uses such as disaster relief (post-earthquake structures), classrooms or small stadia for schools, arts and sports events.

These precycled rental systems could be the future for large sporting events. The pavilion was demounted, and parts of the structure were reused at the Glasgow Commonwealth Games. In the future, to reduce the massive cost to host countries of staging the Olympic Games and the World Cup, it makes more sense if some of the venues are transported to the next country ready to be reassembled. Recycling buildings in this way can help to reduce the embodied carbon needed to construct them.

The impact of statutory requirements

Your design team will advise you on airtightness and how this is critical to reducing your energy bills. The better sealed your building, the more thermally efficient it will be and the less energy you will use to heat it. Airtightness stops heat escaping and cold draughts entering. However, ventilation is a critical part of the design as it prevents condensation and mould growth.

If you are developing a project in a built-up urban area, you should consider a mechanical ventilation and heat recovery (MVHR) system. MVHR will filter and heat the outside air and reduce noise pollution, as open windows and trickle vents are no longer required for ventilation. It extracts the air from the kitchen, bathroom, toilets and utility rooms and takes it through a central heat exchanger. The recovered heat is used in the filtered air supply. This is all achieved through a network of ducts hidden above the ceilings. For this system to work, your building needs to be airtight and well insulated.

If your project is in a rural area, you may prefer not to use an MVHR system as there may already be less noise and plenty of clean fresh air. MVHR will, however, provide a better Standard Assessment Procedure (SAP) and Energy Performance Certificate (EPC) rating and will save energy wherever the building is located. If your building is designed with a well-insulated, airtight envelope with opening windows, you may decide to use trickle vents with intermittent extract or continuous extract (without heat recovery) as your ventilation strategy. You will need to consult your design team to discuss your options to make sure you comply with building regulations.

We have to consider the thermal performance of the fabric as it plays a significant part in determining the building's overall efficiency. The most important value for your project is the U-value, which measures how effective a material is as an insulator. If you are renovating an existing building, we aim to improve the external fabric to a standard higher than that set by building regulations.

The SAP is used for residential buildings and the Simplified Building Energy Modelling (SBEM) calculation is used for all other buildings. These methods are a comprehensive list of measurements to evaluate a building's overall performance. The EPC, the UK's performance rating system for buildings, grades the energy efficiency of a property on a scale of A to G. The EPC is generated by the SAP or SBEM calculations. The methodology is established by the government to demonstrate energy and climate performance.

This is calculated by taking into account how the structure uses and loses energy, based on levels of insulation, windows, renewable energy technologies, types of boilers (heating) and leakage of air. Your engineer will use SAP methodology to analyse design changes and understand how to improve your EPC rating. Improved ratings can be achieved by providing an air or ground source heat pump, biomass boiler, wind turbine, district heat network connection, mechanical ventilation with heat recovery, insulation, photovoltaic panels and solar thermal panels. When a structure is sold, rented or developed, it is legally required to obtain a valid EPC.

If you own a building, you will want to know the impact of future sustainability legislation on your asset. According to Cushman and Wakefield,[14] only 4% of London offices have an EPC which meets the energy use requirements that will be introduced in 2030. This means that without a major overhaul, the majority of London office stock will soon be obsolete.

By 2025, all commercial buildings will need to have an EPC by law and by 2027 if your building does not get a C rating, it will be against the law to lease it. If you do not tackle this issue, then you could be looking at a significant drop in value and diminishing yields from your building as tenants look elsewhere for better accommodation. If you undertake a retrofit project and improve your building's sustainability credentials, there could well be a significant rental uplift – as previously discussed in the Heal's Development case study in Chapter Three.

14 M Phillips, 'You just go, my God': The shocking size of London's green office overhaul (Bisnow, 2021), www.bisnow.com/london/news/sustainability/you-just-go-my-god-the-shocking-size-of-londons-green-office-overhaul-111007, accessed 30 October 2022

The London Energy Transformation Initiative (LETI)[15] has set target goals to increase building performance efficiency and to achieve a zero-carbon future. The aim is for all buildings to achieve a C rating, with the ultimate goal being an A rating in 2030. To achieve these objectives, we need to retrofit our buildings to make them more efficient.

Demolition is not just an ecological issue; it is undesirable for social and economic reasons. In terms of carbon footprint, retrofitting makes sense because of the significant energy savings that can be achieved by converting existing buildings, compared to the high energy costs of demolishing and rebuilding.

There are many benefits to retrofit, both financial and social. Some of the financial benefits include lowering energy bills if buildings are substantially more energy-efficient, and achieving a higher asset value because buildings are more robust and durable. Additionally, occupants' quality of life will improve greatly as they will be living in healthier buildings with decent air quality and thermal comfort.

One of the biggest obstacles to retrofitting existing buildings is a biased VAT system. You will currently pay 20% VAT for most renovation projects and typically 0–5% VAT for new buildings with high embodied carbon. This VAT threshold should be changed to make retrofit projects more viable, and the planning system should be designed to support applications that include retrofit and require a much more robust justification for demolition than is currently required.

---

15 CB George, 'What does net zero mean?' (LETI, 2022), www.leti.uk/netzero, accessed 30 October 2022

A reduced VAT rate of 5% applies where a house has been empty for two years prior to a contractor starting building works. You should speak to a VAT specialist to get advice if you want to take advantage of this. The local authority will require you to confirm the period the property has been vacant for.

Embodied carbon

One of the biggest factors to consider in design is the amount of embodied energy required to construct your building. Embodied carbon is the CO_2 emitted by producing materials through extraction, processing and manufacture; during transportation; and in the construction process when those materials are converted into products, systems and structures.

Disposing of materials at the end of their life cycle also contributes to embodied emissions, so you should always think carefully about which materials you use and make sustainability a priority. During the refurbishment of the Roundhouse Theatre, we retained the brick drum, the cast-iron structure and the existing timber structure, saving over 60% of embodied carbon emissions.

Through design innovation, we can now reduce the amount of steel in a building's infrastructure, resulting in a reduction of up to 20% in embodied carbon. This is also partly due to the decarbonisation of the grid, which means that steel produced on electric arc furnaces will have a low embodied carbon content. To ensure that embodied carbon is kept at a minimum, you will want to know what materials the team is proposing for your building and the amount of energy needed to create and install them. You will also want to know how easy it is to replace and maintain your building, and if it's possible to adapt and recycle it in the future.

To deliver the performance that will be required by the UK government's zero-carbon 2050 targets, and the interim targets that have been set, you should encourage your designers to consider using bio-based materials with low embodied energy levels. We try to achieve this by using as much timber frame construction as possible on our buildings, incorporating Warmcel® recycled cellulose insulation where possible. We also advocate using natural and breathable lime-based plaster systems – especially on renovation projects – together with natural products, such as cork insulation and wood fibre insulation.

It is important to emphasise that not all buildings can be constructed from natural materials and some manufactured products, such as steel and aluminium, may be required. Despite the large amounts of energy it takes to manufacture these products, they are relatively easy to recycle. The recycled content of materials can significantly reduce their embodied energy. The important thing is to make sure that the components in your building – steel frames, for example – are designed so they can be recycled or reused, as we did on the Hyde Park Boating Pavilion.

When implemented in the form of a co-ordinated renovation plan, combining practical and efficient measures results in sustainable structures with a low carbon footprint. This significantly reduces carbon emissions and ensures we are supporting our national transition to net zero. This transition can be assisted by generating renewable energy as locally as possible and replacing fossil fuel heat sources with low-carbon alternatives.

Orientation

One of the most important aspects of sustainability is the orientation of the building. This will determine which rooms receive daylight and sunlight at which times of the day. The amount of sunlight entering a space will determine heat gain and the capacity to store that heat in floors and walls. Any façade facing west or south-west will be subjected to significant amounts of heat. The sun moves lower in the sky as it tracks around to the west, so the screens and roof design need to shade the windows to minimise heat gain.

One of the main issues with large commercial buildings is cooling them down. Most of the year, they are full of people, plant, computers, printers and kitchens that generate heat. One solution is to have exposed concrete walls and stone or precast concrete ceilings, which absorb heat from internal spaces during the day, and openable windows at high level to release heat at night.

Your design team should aim to use natural ventilation as much as possible by orientating the building so that it uses the wind to help cool it down. In summer, if the prevailing winds hit the windward side, negative pressure is created on the leeward side, which helps to suck out the hot air.

If your building is located in an area where there is minimal traffic and pollution, and the floor plates are not too deep, natural ventilation or a combination of mechanical and natural ventilation may be a possibility.

Environmental assessment tools

There may be some value to your organisation in acquiring an accredited certification. The Building Research Establishment's Environmental Assessment Method (BREEAM) is a global sustainability assessment method for master-planning projects, infrastructure and buildings in the UK. It sets standards for building environmental performance at each stage of planning, specification, construction and operation. The BREEAM method can be applied to new developments and refurbishment projects. Your building will be assessed on its sustainability across various categories, such as health and wellbeing, pollution, water and energy. The results indicate how your building performs compared to others in the UK, whether it is in the top 25% of UK buildings with a rating of Very Good or the top 75% with a rating of Pass.

Using the BREEAM assessment tool is useful as it allows you to assess the value of your asset over its entire life cycle and helps give focus to and guide the design of the project. The assessment is, however, expensive; it involves engaging a BREEAM consultant and significantly increases the workload for your consultant team. You should carefully consider why you want to do it and how the results will benefit your organisation. Will it help you to design a better building or could the money be better spent elsewhere on other specialist design input?

Summary

There is no doubt that we are facing an energy crisis as a result of a warming climate and the loss of habitats and biodiversity. You can play an important role in reducing your energy use and emissions by commissioning consultants who share your values and want to make a difference. A sustainable design will reduce your energy bills, improve the wellbeing and health of your employees and clients and enhance your brand.

In the UK we are fortunate to be part of a long tradition of high-quality research in environmental design. To achieve a remarkable building or masterplan, your design team need to be well integrated and work together from the earliest stages. As demonstrated by the Masonic Lodge in Harrow, it is when environmental engineering is combined with architecture that remarkable solutions emerge.

Retrofitting your existing buildings to meet sustainability targets and new emissions standards is going to be a major part of upgrading and future-proofing your projects. This work requires attention to detail, invasive investigations and a dedicated team to overcome the inevitable tension between energy-saving and heritage.

All buildings should be designed by thinking about environmental and structural systems together from the beginning. These ideas should drive how the building form develops in conjunction with your brief. If MMC can be incorporated into the building design, the project will be more precise and more efficient. This will reduce the building's embodied carbon.

Sustainability is about getting the design and orientation of the building correct so you can take advantage of the sun and wind and prevent your building from overheating. There has never been a more urgent time for leaders to show the way to a brighter, more sustainable future.

CHECKLIST

- ⭕ High-quality natural light and air will improve the productivity of your staff, pupils and members.

- ⭕ Commission a masterplan for your school or club that includes new buildings and extensions, and changes to existing buildings, future infrastructure and energy use.

- ⭕ Sustainability will help build your brand and attract more people.

- ⭕ In 2027 you will not be able to rent your building if you don't have a C rating on your EPC.

- ⭕ You must consider the orientation of your building to prevent overheating on the western side and to take advantage of solar energy in the winter.

- ⭕ Efficient and precise design using prefabrication can help make your project more sustainable by reducing waste and construction time.

- ⭕ Consider during the early design phase if your building (or parts of your building) could be easily adapted, recycled or relocated in the future.

- ⭕ Consider heat pump technology to future-proof your building and reduce your energy costs.

ACTION STEP

To start thinking about sustainability, take our 'How prepared are you to meet sustainability targets on your building renovation?' scorecard.

You can find this at
www.grahamfordarchitects.com/scorecards

Conclusion

This book is a guide to help you develop a successful strategy before a project begins. It explains how you should organise your masterplan or building project, the principles and the steps you need to go through to make sure of a smooth design and construction process, and how to approach planning permission. Your goal should be to create a sustainable environment that utilises your existing assets and creates better spaces and access to your school or club. You will want a facility that costs less to run, is sustainable and attracts parents and pupils, tenants or new members and at the same time enhances your brand. It is worth remembering that if you are commissioning a new masterplan and new buildings or renovating existing buildings, you will be making a significant contribution to the country's cultural inheritance that will be enjoyed for generations to come.

We started by looking at the critical importance of writing a good brief for the design team. Your architect or design advisor needs to understand how your organisation works and what your strategic vision is for the next ten to fifteen years. They can then use their spatial skills to work with your team and your project manager to create a physical masterplan and designs for buildings and landscape. The experience of designing a masterplan with new buildings should be a rewarding one, in which many combined intelligences are harnessed for the benefit of your project and your considerations are put at the centre of design and construction activity.

Your vision will be driven partly by your values and partly by your customers' needs, interests and motivations. The masterplan is a tool that uses data, and looks at numbers of staff, pupils and members to plan how your organisation can become future-proof. Other factors will include the impacts of artificial intelligence, energy supply and the digital future.

A big challenge you are likely to face on your journey is gaining planning permission for your development. To help you navigate this, you will need to commission a great team that have worked together in the past. The team will help you develop a strong value proposition that emphasises the social, economic and environmental values in your scheme. The planning department's role is to assess the impact of your proposals on the local area and the wider built environment. They will review your scheme against their local policies and the city plan. The stronger your story, the more chance there is your project will be approved. Once you have overcome the planning hurdle, the value of your site will have substantially increased.

You should discuss what procurement route to use with your project manager, architect or design advisor at the early stages of the briefing process. No matter which path you take, it is important that you don't select a contractor purely on the basis of cost. If you select a design and build contract, it would be wise to ensure your employer's agent's requirements are robust and well developed and that you have mitigated all the risks on your site. It would also be useful to review the developing contractor's design with your team so you can assess the quality of the work they are producing before construction begins.

Your architect must be a great manager who can work in different roles to help you on your journey. I have described the different roles I have worked in to help my clients with planning, value propositions, alternative design solutions and problem-solving. Your architect will have their own unique set of skills and tools they can use.

I have introduced you to my SPECS principles: the essential ingredients in the recipe used to make your project. Without these ingredients, your development is less likely to be a success. **Simplicity** is important in the early design stage and captures your vision statement for the project. Your architect will respond to your vision and early brief and produce a diagram of the new building or the masterplan to test ideas. This diagram will contain early ideas of how to integrate structural, mechanical and electrical principles that support the architectural concept. **Precision** is important when you develop your brief with your team and describe in some detail what your negotiable and non-negotiable items are, and it is important in all project stages to make sure you build up layers of information that are accurate so the original drawings can eventually be developed into construction drawings. **Efficiency** is important to ensure that the building process maximises both speed and quality, and that logistical issues around your site are considered at the beginning. Efficiency will reduce disruption, and meaningful **collaboration** is crucial to a successful project. A team approach is the only way to close the gap between abstract plans and the reality of the construction site. **Sustainability** is a key component of your value proposition and will help you attract better tenants for your building, members to your club or pupils to your school.

If you are renovating an existing building, you will need to upgrade the fabric and the ventilation systems to make sure your asset does not diminish in value over time and you can meet new statutory regulations.

Do not be tempted to cut out any of the nine steps I outlined in the process – it could put your planning application at risk due to the costs of resubmission or it could mean a compromised building when completed. A quality building will ensure you are able to compete with other schools or clubs who are innovating and investing.

Finally, make sure you have a strong team around you who can help identify and mitigate risks and leverage the spatial intelligence and the technical skills of your design team so they can deliver great value. This will help to make sure the bumpy road of construction is smoothed out and everyone is able to deliver what you need for the project to be a success.

Acknowledgements

This book is dedicated to the team at GFA architects who contributed to the design of all the projects, especially Simona Roggero who has worked with me for over six years on all the projects referred to in this book.

I would like to thank Roger Whiteman from Beside Design, Alan Arnott of Studio 9 and Martha Wailes, who provided valuable feedback on early versions of the text.

I would also like to thank Daniel and Andrew Priestley from Dent Global who helped me to identify the principles which I believe are critical to the success of all our projects and galvanised me to write this book.

The Author

Graham Ford was educated in New Zealand, the United States and Spain. He has been the recipient of numerous awards, including a Victoria University Master's Scholarship and a Todd Foundation Scholarship. Graham has a postgraduate degree in sustainable architecture and a doctorate in practice-based research.

Graham has twenty-five years' experience in professional practice and founded GFA in 2006. Since 1996, Graham has worked for and collaborated with some of the UK's leading design consultancies, including Wilkinson Eyre Architects, Landolt and Brown Architects and John McAslan and Partners. Graham has taught, lectured and been a design critic at Cardiff, Nottingham and Reading Universities. He is currently teaching part time at the School of Architecture at the University of Portsmouth.

Graham has practical experience in the application of the principles of sustainable design to buildings in royal and regional parks, green belts and in conservation areas. His work focuses on sustainability as an integral part of architectural design.

- www.linkedin.com/in/graham-ford-2144746
- www.facebook.com/grahamfordarchitects
- @gfordarchitects

www.ingramcontent.com/pod-product-compliance
Lightning Source LLC
Chambersburg PA
CBHW010045090426
42735CB00020B/3396